HORSE CRAZY

The Story of a Woman and a
World in Love with an Animal

Sarah Maslin Nir

SIMON & SCHUSTER

NEW YORK   LONDON   TORONTO   SYDNEY   NEW DELHI

Simon & Schuster
1230 Avenue of the Americas
New York, NY 10020

First Simon & Schuster hardcover edition August 2020

SIMON & SCHUSTER and colophon are registered trademarks of Simon & Schuster, Inc.

For information about special discounts for bulk purchases, please contact Simon & Schuster Special Sales at 1-866-506-1949 or business@simonandschuster.com.

The Simon & Schuster Speakers Bureau can bring authors to your live event. For more information or to book an event contact the Simon & Schuster Speakers Bureau at 1-866-248-3049 or visit our website at www.simonspeakers.com.

Interior design by Carly Loman
Photo on page ii by Diana Zadarla

Manufactured in the United States of America

10 9 8 7 6 5 4 3 2 1

Library of Congress Cataloging-in-Publication Data

Names: Nir, Sarah Maslin, 1983- author.
Title: Horse crazy / Sarah Maslin Nir.
Description: First Simon & Schuster hardcover edition. |
New York : Simon and Schuster, 2020. |
Identifiers: LCCN 2019058615 (print) | LCCN 2019058616 (ebook) |
ISBN 9781501196232 (hardcover) | ISBN 9781501196249 (ebook)
Subjects: LCSH: Nir, Sarah Maslin, 1983- | Horsemen and horsewomen—United States—Biography. | Horse owners—United States—Biography. | Women journalists—United States—Biography. | Horsemanship—Social aspects—United States. | Human-animal relationships.
Classification: LCC SF284.52.N57 A3 2020 (print) | LCC SF284.52.N57 (ebook) | DDC 798.092 [B]—dc23
LC record available at https://lccn.loc.gov/2019058615

ISBN 978-1-5011-9623-2
ISBN 978-1-5011-9624-9 (ebook)

To all the horses I've loved,
Amigo, Willow, Trendy, Bravo, Stellar,
and every single one I've ever set eyes on.

HORSE CRAZY

INTRODUCTION

In the decade I've worked for the *New York Times*, I've reported across the country and around the world. And as soon as I file each story, I do one thing before I head home: I search for the horses. The rider in me wants to gaze at them, stroke them, gallop with them, but the reporter in me has only one goal: to know their stories.

And so I've found myself, notebook in hand, interviewing the keepers of the street horses of Senegal, West Africa, as the animals slept in corrals of parked cars. I've traced the Viking history of the canny Norwegian Fjord horse who extracted us both from a peat bog in the Scottish Highlands. And I've quizzed Indian soldiers about the indigenous battle horses I charged through a quarry in Rajasthan.

For my entire life, I've sought out horses endlessly, even in the urban world in which I grew up. As a girl, I found them hidden between the townhouses of Manhattan's Upper West Side, underneath the Robert F. Kennedy Bridge in Harlem, and stampeding through Central Park.

And yet, all this time, I never asked myself why I love horses.

That's because the answer has always been *because horses*. It's a response that anybody who has ever felt the ineluctable tug of their big amber eyes, in which you see something much more than your own reflection, or who knows the peace of their breathing, and the shattering wildness of their gallop, immediately understands.

Because horses. Answer enough.

But while to me horses feel like an inevitability, a part of my body and life in a way I don't question any more than I would the rise and fall of my own chest, the reporter in me is plagued by and duty-bound to ask. In fact, "Why?" is the sum total of my job description. So it was only a matter of time before I turned the query on myself.

That quest became this book, and expanded far beyond just me, because I'm not alone. As I sought out the horses, I found their humans. The two teen sisters who bought a wild pony only to set her forever free. The executive who left corporate America for a life patrolling Central Park as a mounted ranger. The Pennsylvania man who ran an equine version of the FBI. The fox hunter who galloped away from a crumbling marriage. The diplomat's stepdaughter who wanted forbidden horses so badly she smuggled their semen across the sea.

"Horses lend themselves to stories," I once wrote in the *Times*. In the United States in particular, horses, their manes streaming, nostrils flaring, hooves thudding, carry with them something of our projected national psyche. There are over 7 million horses in America, far more than even when they were our only way to get around. They are not necessary at all, yet for many they seem more so. Here, they are furls of an American flag in equid form, imbued with our narratives of national identity. They carry on their backs the tales we tell ourselves about who we are.

Who I am.

When I finally asked myself why, the search grew epic. I dug for the reason in the belly of a 747, huddling with a trio of Dutch warmbloods as we crossed the Atlantic in the cargo hold. I looked for it floating in the dawn waters off the coast of Virginia as all around me ponies swam in the salt. I sought it underneath their thick lips and listened for it in the syncopation of their seven-league strides. In the science behind their piano-key teeth and the music of their bugling whinnies.

There are theories: a horse's stride replicates an essential rhythm we all felt for those first nine months of our lives, rocking within amnion as our mother went about the world. Another posits power: horses lend us their own, extend our feeble human legs with their muscled limbs. They allow us to tap into their strength and seize it as ours, to feel speed and might far beyond our capacity, to touch something close to the infinite. On my own two legs, I'm just Sarah. Lent four more, I'm formidable.

Both theories seem right—yet both don't come close to capturing what it is I feel when a horse snuffles my palm or graces me with the perfect jump, or simply stands still in a paddock in beauty that hurts. At last, I realized that the only one capable of answering my question would be those who know me best in this world: the horses themselves. *Horse Crazy* is structured around the lifelong dialogues I've had with these animals, each chapter named after a horse who told me its story or helped write my own.

I don't believe that riding a horse is a dance, as some say, with a partner who enjoys the experience the same way I do.

Instead, I believe it is a conversation, an intimate dialogue between a creature that over the millennia has become a perfect foil, partner, and complement to humankind. To me.

I realized I've been having that conversation all my life. Here, I'll tell you what I've heard.

GUERNSEY

I don't remember the first time I was on a horse. I remember the first time I was off one. I recall the shape of the rocks in front of my eyes, my scraped cheek pressed to the ground. Those stones come to me in a silhouette of massive, misshapen pebbles. They loom large, like a city skyline gone sideways.

I was two years old. I still can see the stones.

I was perhaps all of three feet tall, that little length of me now stretched out in a rocky field. In the distance was an imposing brick mansion, and on its rolling grounds, horses idly cropped the grass. Perhaps one or two raised their heads from the ryegrass at the commotion, decided the plight of a little child splattered in the dirt wasn't their problem, and went back to chew. I wouldn't know: I just lay there on the ground, cold pebbles pressed into my face, rocks turned mountains that close to my pupils—and stayed there.

In truth, I was fine, bruised, with just a tear in my teeny jodhpurs, despite the fact that I had just plopped off the back of a full-sized horse, but I didn't know that at the time. What I knew was the shock of a perfect day gone upside down. It was my third-ever riding lesson at this imposing, horse-dotted estate in Amagansett, New York. I'd arrived that morning, bold and begging to do more than plod along. I was now regretting it.

The horse was called Guernsey, a tremendous patchwork of white and latte-colored spots like the dairy cow breed after which he was named. I had brushed him to sparkling and watched as the tiny mishmashed contraption I used as a saddle was tethered to his back. At the mounting block, I'd waited patiently for a grown-up to lift me onto his back. Out in the field, my teacher clipped a lead to his bridle, and we began our promenade in a circle.

I was bold on a horse, as tall at last as I felt inside, as powerful and in control as a child is not in any other part of her life. With a kick from my little paddock-booted feet, he jogged into gear. I'd barely mastered the trot by this early point in my equestrian education, but I was *sure* I was a cowgirl anyhow. I jammed my small heel on the gas—his sides—and he lumbered into a floppy canter. We were off!

But then, suddenly, *I* was off, grit under my tongue, a rider without a horse. Had I looked up, I would have seen him. My mother, at the fence line in her Hamptons camouflage of pastel-popped collar and white jeans, did. I knew that because she was screaming. That's because the horse, ever so obediently, was still cantering around the circle. I lay in his path. He was barreling down on me.

I began riding at age two, and the family lore is that it was because I would never sit still. Car rides from Manhattan's Upper East Side where we lived out to East Hampton where my family had a summer home were apparently torture for my parents. A

Ford Taurus filled with my old dad, fifty-five years of age with no time for the antics of a fourth and final child, and my mom, thirty-eight years old and overwhelmed by kinetic little me. My father was a psychiatrist, and I arrived late in his life. Born in Poland in 1930, he was old enough that he was a Holocaust survivor who had evaded Hitler's grip as a nine-year-old boy. Dad had two sons from his previous marriage, and by the time it was my turn with him as a father, his days of babbling at babies and humoring toddlers were long gone by. Although he was a renowned child psychiatrist, he was better at relating to little ones as patients than as his progeny eager for him to get down on the floor and finger-paint.

My parents met when my mom was an assistant elementary school teacher and my father had been called in to diagnose a diabetic student of hers who seemed suicidal: he was refusing his insulin. My father cracked the case with his discovery that the child had absorbed the messages of back-to-school specials a little too well; the child told Dad he was afraid that if he injected the lifesaving drug into his veins, he would become "a druggie."

My mother was impressed. Detecting in the visiting doctor's voice the lilt of an Israeli accent, she ran down the school staircase after him as he left the session, shouting the formal Hebrew-school Hebrew she knew. "*Adoni!*" she called out, thinking it just meant "sir!" (In fact, the nuanced translation is more like "my liege.") My father, then forty-two years old, bald, mustachioed, with teeth snaggled by the wartime blight of his childhood in Poland, stopped in his tracks. He looked up the staircase and saw a green-eyed, twenty-five-year-old woman with waist-length blond hair.

And she had called him her lord.

In 1980, nearly a decade into what ended up being a forty-one-year marriage, they were two doctors seeking to climb a ladder of affluence. To do so, they had bought a 900-square-foot shack in East Hampton, in an unfashionable area of fishermen and mechanics called the Springs, far from the glamorous hordes. The floors of the shack were covered in rust-colored shag carpet and peeling linoleum, but it was set on a glorious cliff overlooking Gardiner's Bay. Over the years, as they ascended the ranks of success and prestige, the house grew. Toddler me didn't notice the cramped quarters because I was almost never inside. My little life was with the hermit crabs that played out funny pageants only they could understand in the tide pools that eddied out back. And with the horses I sought out endlessly in the woods and fields between the shingle-style mansions.

It was a quainter time when such things like a beach house and even our Park Avenue apartment were somewhat accessible to those who were working but not well-off. My parents bought the sprawling Upper East Side home where I grew up in the late 1970s for $45,000, less than some cars costs today. It was a time when New York City was teetering toward bankruptcy, and people were fleeing their Classic Sixes in droves. So many people had defaulted on purchasing this apartment before them—scared of the harbingers of a city spiraling toward decline: an economic recession, the blackout riot the summer of '77 and spiking murder rates—that they had to pay with a Mr. Monopoly–style bag of cash. They borrowed $10,000 from my maternal grandmother, Frieda, a secretary, the apartment a rope they climbed into the middle class.

Inlaid in the parquet floor under the dining room table when they purchased the resplendent apartment was a tiny buzzer. Scrabbling up from working-class roots, my family needed the doorman to explain what it was for: the lady of the house pressed it with her toe, where it would ring in the kitchen and summon her staff.

The first thing my mother did when she moved in was rip out the wiring.

On weekends, we went to the beach. Driving down Old Stone Highway in Amagansett, the next town over, in the Taurus one afternoon, we passed a spotted pony. She was nut-brown and splotched with white, a pattern that is known in the horse world as a pinto, and stood in a small pen by the side of the road. The pen was overgrown with vines of trumpet creeper so that she stood chest deep in pink petals. That's when the epiphany hit my parents: putting me on a moving horse would be the secret to getting me to sit still. That is, I'd be moving but seated, rooted to where they could see me. On a horse, I could be as hyper as I itched to be but unable to skitter out of sight. They turned up the vine-laced drive.

They had no idea what their clever plan would set in motion.

The barn was on a private estate owned, I was told, by a textile magnate, who had converted the rooms in the mansion into "showrooms" for various rugs. He had three horses of his own that he barely rode, except to jump on their backs every so often, startling them out of semiretirement to prove to himself he was a gentleman farmer. Occasionally I'd see him puttering past the paddocks on a farm vehicle, back and forth, for no discernible purpose, but with all the ceremony of the pope in his popemobile.

I didn't ride the horse called Guernsey at first, or the spotted pony, either. Technically, no riding lessons were offered on the estate, on which a handful of privately owned horses and ponies were boarded. But the owner of the pony, which had the swoon-inducing name Cutie, let my mom leave me there for an hour or so some days to play with her waist-high pet. Like all other personality-packed ponies, Cutie was the paradigmatic example of what I'll call Sarah's Axiom of Ponyness: a pony's troublesomeness is in equal proportion to how sugar-sweet adorable it looks and in inverse relation to its height. In plainspeak, the teenier the pony, the ruder. A bite-sized one named Cutie? Watch out.

Cutie's owner had decided the pony was either too opinionated or, more likely, too tiny to ride, so my first equine experience was not riding but clip-clopping down Town Lane with her owner as the pony towed us in a tiny wagon. Grandma Frieda took me to be with Cutie daily. She and I lived together out at the beach all summer while my parents saw patients and plays in the city. They left me Out East, as Manhattanites refer to the eastern end of Long Island, with my maternal grandmother. Grandma tried to be both mom and dad for me while I pined for parents who had little time for someone as yet too young to make cocktail conversation.

Grandma was barely five feet tall, her skin speckled with vitiligo, an ebbing of the pigmentation that made her hands as bone-white as they were soft. She never married after Grandpa David died of a heart attack when Mom was twenty-five. Her sole pleasure of the flesh, it seemed to me, was a small bowl of coffee ice cream each night while mooning after Alex Trebek on *Jeopardy!*

Her softness belied a hard life. She had given birth to two children who died just hours into this world. Yet she seemed to see her suffering as trials on the path to joy. Through that loss, she found my mother and my uncle, she would remind me, whom she adopted as newborns. The only time it dawned on me that she and I were not blood-related was when as a teenager, I shot up to five feet eight, towering over all my petite family at bar and bat mitzvahs.

"Thank God you're not a shrimp like the rest of our family!" Grandma said when I told her I felt out of place. "Shrimp aren't kosher!"

She was so tolerant, so genial, that it became an absurdist family joke on long car rides: we would break up the tedium of the drive by warning my silent, smiling grandma that we would have no choice but to tow her behind the car to East Hampton on roller skates if she didn't stop bellyaching.

In the city, Grandma wore skirt suits of the *Golden Girls* variety, purchased at a proper store for women of a certain age. Out East, as close as she ever got to casual wear or trousers, was a skort my mom gave her in an effort to make her hip. Even so, when I became a tween, she was the first to take me shopping for tube tops and miniskirts, applauding as I modeled postage-stamp-sized outfits for her in the dressing room.

Grandma wasn't old-fashioned or retro (she wore a rainbow gay pride pin on her pocketbook when I asked her to in support of my best childhood friend); she was just a grandma to her core—so much so that the doormen at my family home in New York City called her "Grandma" too, and she wouldn't respond to anything else.

Once, when I was just over a year old and we were together

alone in East Hampton, I crept between the railings of the porch and tumbled off. Grandma drove me to Southampton Hospital shoeless, in her nightgown. I loved hearing the story. Later, I realized why: the image of my tiny, dignified grandma so undone, braless and barefoot, and heedless of anything other than the need to protect me was the first time I felt important to anyone in my life.

Grandma Frieda was my confidante and chauffeur, driving me to the barn each morning. At bedtime, she tunelessly sang me nursery rhymes without cease, holding one of my hands as I sucked the other's thumb, and not minding when I swapped a sticky hand for a fresh one. I felt so alone I could only fall asleep gripping her to make sure she couldn't leave me too.

I always made her repeat one song in particular until I fell asleep.

I had a little pony, her name was Dapple Gray,
I lent her to a Lady, to ride a mile away
She whipped her and she lashed her!
And she rode her through the mire!
I will not lend my pony, for a lady's hire.

Sometimes in the pony cart, Cutie's owner would allow me to steer, but at first, we didn't get far. I'd taken a personal oath to never whip and lash Cutie like the lady in the poem, and I refused to so much as flap the reins over her spotted rump, so of course we did not move. Instead, I learned to click my tongue. At the sound, Cutie would tug at the traces and march us down the road.

But idyllic as it was, it wasn't riding. Across the street was another barn, a commercial riding operation, and I begged my mother to take me there one afternoon after clopping down the roads with Cutie, so I could sit on an actual horse and ride. Unfortunately, you had to be five years old to take a riding lesson, they said; two was far too small. It was an agonizing fact that I bawled over as we left, hammering at the window of our station wagon and begging whomever was up there listening for three years to pass by the time we drove home.

To silence the squall, my mother pulled back into Cutie's estate, and accosted a willowy woman she found there astride a huge pinto horse; that was when I met Guernsey. His owner's name was Diana Zadarla, and she was a babe and a bohemian rhapsody—a white woman who wore her hair in Indonesian wraps. She was a thirty-something artist who worked at a local gallery to afford her other art, riding, and ducked out on her lunch break to teach me on her horse. She had a heart-shaped face and soft eyes that crinkled, and some part of me knew instantly she loved horses with a fierceness that I was finding in myself. Diana didn't give lessons, but somehow that afternoon, my brash New Yorker mother tawked Diana into setting me astride her full-sized horse. That day and every day after.

Guernsey was the first animal I ever sat on. I reveled in attempting to brush his white patches until they shone and polishing his candy-striped hooves. I didn't know it then, but he was common and coarse. Instead, I saw just his two starry blue eyes. Beneath his divine eyes, he had a porcine, pigment-less pink snout; Diana would let me rub sunblock on it whenever we rode, lest it fry.

Being a Hamptons horse, housed on the grounds of a red-brick mansion, Guernsey would later have singular distinction of teaching at least one celebrity how to ride. Michael J. Fox mounted up to prepare for an iteration of *Back to the Future*, where he travels back in time to the American West, something Diana was proud to share with anyone who would listen. As a child, I thought an actual fox had ridden Guernsey. The horse was so tolerant that it made total sense, and so I never questioned my misunderstanding of the boast.

Even if he had never been sprinkled with stardom, Guernsey looms in my mind. In fact, he was mammoth. Unlike people, horse height is not measured to the top of their head, which will bob up to the barn rafters at a rustle in the hay or dip low for clover in a gully, and is thus an unreliable read on just how tall they are. They are measured instead to their withers, the bony spot formed by the meeting of their shoulder blades at the base of the neck.

The withers is a useful point of reference because no matter how the horse moves across the ground, this measurement does not change. And in a way, it never has.

There is a wooden rod that is at least 3,350 years old that rests in a temperature-controlled case in the Louvre museum in Paris. Thin and beveled along one side, the circa 1136 BCE doodad belonged to an ancient Egyptian bureaucrat named Maya, a treasurer under Tutankhamun. Etched all along the side are teeny hieroglyphic images of fingers, palms, and hands. It is a ruler that, luckily for archaeologists, bears along its length a glossary of ancient units of measurement. Like the Rosetta Stone was for language, this code-cracking cubit rod is for metrics. In that ancient Egyptian metric system, a finger correlates roughly to an inch. Put your fingers side by side together and measure them across, and you get about four inches, a hand's span. That four-inch "hand" is a unit of measurement long slammed in a sarcophagus and left to history.

And yet, that antique unit is still used to measure all horses, on racetracks, in fields, or in show jumping barns today, just as it was in the kingdom of King Tut. (It is not the only ancient esoterica that persists; a dusty attic full of relics lives on among horses and their people. The convention of mounting on the left side is another example: it comes from cavalry, who got on their horses from the left so as not to accidentally stab them with their sword.)

Hands are used to splice equines into categories by height. Ponies like the not-so-Cutie, are generally the animals that come in under fourteen and a half hands from their hoof to the bony peak of their shoulders. Horses are anything higher.

In the pink-creeper-covered stables in Amagansett, Guernsey seemed a thousand hands high.

Guernsey was as big as I was small, so we had a problem: there was no saddle simultaneously large enough for him and teensy enough for me. Diana's artistic creativity came to the rescue. She took a square of cloth and sewed on tiny steel stirrups. With a jerry-rigged system of breastplates and girth, she affixed the patch of cotton to his back and me atop it.

Blame the patchwork "saddle," blame the fact that truly, two years old is too small to start an extreme sport, or blame the fact that my cow-colored behemoth was just too large, but the first time Guernsey picked up that lumbering canter, I pitched headlong into the dirt.

And so, back to where we began this story: two-year-old me, the ground, the rocks, and Guernsey barreling down on me while everyone gasped.

And, ludicrously, I just lay there, looking at the gravel. That is my first memory of horses.

And of the grown-ups screaming. That afternoon, I had been riding Guernsey as he was affixed to what is called a longe line, a long lead tethered to the bridle and held by a trainer who stands at the center of a circle, commanding the horse in a roundelay on the perimeter. It's a useful training tool for a beginner, allowing the rider to focus on balance and pace and not worry about steering. Restrained, the animal loops on a track as certain as the carousel horse bobbing on a painted pole.

But the nature of a circle is, of course, that it circles back around. And so as I lay flopped there, Guernsey completed the revolution, and he swung back full blast toward me. Unburdened of his rider, dutiful, trusty, gigantic Guernsey kept doing his job; he kept thundering toward me. My mother shrieked. I was about to be equine roadkill. But shouts don't often make a horse stop. More often, a loud yelp will startle a horse into charging harder. Right at me.

Then, with a whoosh, the little lump of two-year-old on the dirt that afternoon lived. Guernsey had jumped over me.

And that's it really. That's the whole thing of horses, the ones who leap over you and spare you and indelibly print their silhouette against the sun. Those who remain forever the hawk-like shadow suspended overhead that afternoon. Even the ones who break and bruise you, but so, so beautifully. That's the gift of a horse, the thing we take from them and never really return. We take pleasure in their life, find freedom in their hoofbeats.

They stir me, but I often wonder, Do we do more than succor them in return? Does anything make a horse feel the way a horse makes me feel?

"Guernsey," I remember saying as I rolled over and stood up. "Good horse."

On a morning after a thirty-two-year absence, I emailed Diana, his owner. I had ridden Guernsey for a few more summers until I was old enough to make the cut for a local pony camp. We had lost touch beyond the occasional carrot I stopped by to deliver to my old mount. She was nearly seventy and a sculptor with a whimsical business, scraping three-dimensional horse portraits out of clay for a living. Her batik-wrapped hair (still, I imagined as I read her emailed response, and I don't want to know if it isn't—some things just aren't allowed to change), she told me, is now silvered in waves.

She responded to my emailed hello with a picture. In un-Instagrammed sepia—the actual color of 1985—there is Guernsey, sensitive blue eyes closed gingerly against bright summer sun. There is Diana, her hair piled under swirled cloth. And there, high on his back, am I. Memory had not failed. I was, it turns out, roughly as big as Guernsey's head.

Diana's phone number in East Hampton was in the email. "Sarah! I've been waiting for your call, just a moment!" Diana set the receiver down in her house on Red Dirt Road, and I could hear the rustling of paper. "I've kept it all these years," she said when she picked up the line again. In her hands was a preserved slip of yellowed newsprint that she had long kept in a manila file in her home and thumbed through from time to time. It was a clipping from the local newspaper, the *East Hampton Star*, saved since 1995: Guernsey's obituary.

"Guernsey could be a strong ride for an adult, but he possessed a rare and wonderful personality that made him, despite his formidable strength, the gentlest and kindest of creatures with children," Diana read in a clear voice that I remembered from when it commanded me to put my heels down and eyes up across the courtyard of the creeper-covered stables so long ago.

" 'Guernsey' is one of the first words several generations of children have learned to utter," Diana read. *Guernsey*, I thought. *Good horse.*

He wasn't always so gentle, Diana told me. He began life roughly, somewhere out West. But he had softened with time and, perhaps, cushy Hamptons living, a town he arrived at as a tumbledown three-year-old in 1973. "He bore the scars of a wild western youth—a pierced nose, knife-slashed tongue, and a brand," the obituary continued. "As he traveled along the roadsides of Amagansett, people would pass in automobiles and call his name."

She read to me the last line. It doubled as his epitaph—and perhaps that of all the horses I have ever known: "That a single animal could give so many people and other animals so much pleasure over the course of his lifetime is a blessing to us all."

We sat quietly, me in New York City, Diana on the small farm where she lived. Then she continued, not reading now. "I'm sixty-eight now, and I'm still galloping through the woods, and people are horrified when they hear this," Diana said. The receiver to her ear, she was watching her two draft-crosses as they sniffed through scattered fall leaves. Jefferson is mostly Suffolk Punch, an English breed, with a stippled brown hide of silver dollar–sized dapples across a rump like a wine cask. Sir Oliver is a silvered part Percheron. In the evenings, she lets the huge, lumbering horses out of their paddock to trim the grass in her front lawn, nipping the edges of her driveway under the cover of dusk. Two jiggly basset hounds run low around their heels.

"Twelve. That's my age with horses," Diana continued. "That was when I had the fantasy that they would take care of me, even though I know now that's not true—I've suffered enough for me to know better now." She sighed. I heard the newsprint clipping of Guernsey's obituary crinkle in her hands.

"But that kind of free, fearless thing that I used to have at that age—which God knows I don't have now—it clicks in every once in a while," Diana continued, her voice brightening to that same sunny tone that told me to brush off the dirt and get back on her giant Guernsey that summer day when I was two years old. To mount once again her two-toned beast with the blue blue eyes.

"I feel it only one way," she said. "When I'm on them."

KOŃ

/kɔɲ/ *Polish*: a horse

"We're horse people!" my dad said roundly one afternoon when I was about seven years old. Back then, he smoked cigars, truly committing to the antique Viennese vision of the crusty smoke-shrouded psychiatrist. I liked it, because he would give me the thin paper rings from around them, and I would put them on my fingers as he puffed and pretend I was betrothed to some Cuban-loving knight. I sat at his feet. My thirteen-year-old brother, David, squawked his clarinet somewhere in the distant confines of our country house. "They've been in our family for generations!"

My ears pricked. Even at that age I understood and loathed that my entrée into the sport was as an outsider. Everything about me was, a way of being in the world inculcated into me by family lore, by the narratives that tethered and constricted like sinews running taut through my life.

Externally, I appeared every bit part of the life my parents had devised for me, but that never occurred to me for the long years of my youth. I felt like an interloper, a spy, in my elite private school, Brearley, where it seemed I was the only one out of the 656 girls who brought kosher lunch meat on field trips and asked in the cafeteria if the soup contained pork. I felt like an outsider even as my address was 1050 Park Avenue because my mother was born out of wedlock, illegitimate issue of an illicit rendezvous of an Irish nurse and a Jewish doctor. She was abandoned by them, given up for adoption to my grandpa and grandma. Grandpa David and Grandma Frieda, the offspring of immigrant Russian Jews found themselves the instant parents of a green-eyed, flaxen-haired babe. Her narrative of abandonment, of being a stranger in a strange land, interlaced with my own.

But mostly it was because even in my plush life, it felt like we were still in hiding, so crisply is trauma transmitted through generations. My father's early experience of being concealed in plain sight from the Nazis somehow felt to me that it continued on Park Avenue. I hoped our lavish address was the ultimate armor. Who could rip us from our lives again when we presided over the turret of the castle of the world?

Sometimes I woke up nights in my room in the back of the kitchen, worried the Gestapo—a word I had so often overheard while playing with plastic horses under the dining room table that to me it just meant boogeyman—had come. Other times I was afraid to explain to the blonde and barretted competitors in the short-stirrup, or kiddie, arena that I had been absent from a competition because it fell on Yom Kippur. I had muddled in my baby mind that their Aryan phenotype meant they were actual Nazis. There is a joke in my family that you can't have a meal finish without someone mentioning the Holocaust: sometimes when no one has brought it up yet and dessert is scraped clean, someone will yell "Holocaust!" and we will laugh and push out our chairs and leave the table.

Looking back, it's not very funny.

I felt like an outsider because my dad was old and didn't know the rules of baseball. He was emphatically a foreigner. When he moved to America, he arrived at his first Fourth of July party dressed in a tuxedo because he had assumed that the celebration of the birth of the nation was an occasion that called for formal wear. And where American dads watched baseball, my father's spectator sport was opera.

Second only to his love of Giuseppe Verdi, Giacomo Puccini, and Georges Bizet was his love of bragging about how little he paid for a seat to hear opera. He'd go solo to Lincoln Center most weeknights in New York's winter. There, he'd hang out by the dancing fountain at the center of the plaza and try to spot the lovelorn—those who'd been stood up by opera dates and had a ticket to sell. He would approach them only minutes before the curtain rose. The seller would suggest $100; my father would hold up a crumpled $20. A few moments later, Dad would usually be snug (and smug) in the front-row velvet by the time the orchestra raised their quivering bows.

My father's favorite aria is from Verdi's Aida: "*Ritorna Vincitor*." Return of the victor. Dad viewed his successful life as a magnificent victory lap, but I viewed it as tenuous. The success my parents had both amassed, despite their brutal beginnings, was not truly ours, I felt. It all seemed contingent, ephemeral, and liable to vanish. Just like my father's bourgeoisie life had when the Nazis invaded and murdered my grandpa. Just like my mother's biological parents had themselves vanished. How could I possibly belong to my family's new life?

I think about why I chose horses to devote my life to, and I think of the soft muzzles and limpid eyes and thrumming heartbeats that so draw me to these animals. But trained by my Freudian father, I can't help but think harder and unpack all of what equestrian sport represents in my society. It is the sport of kings and Kennedys, a pursuit dripping with elitism and Americana. As the progeny of immigrants, of people who did not belong to this land, I was claiming rights to the leisure of the Other. "Ralph Lauren was born a Jewish boychick from the Bronx named Ralphie Lifshitz!" my dad would tell anyone who would listen, and indeed it is true. Ralph understood my need to take cover, to escape the *shtetl,* or Jewish ghetto, for the safety of the *ubermensch,* to camouflage in their cashmere and jodhpurs.

So when Dad casually tossed out the fact that our family were horse people that summer day, my heart leaped. Dad had a string of catchwords and phrases he used ad infinitum, both in conversation and in his practice where he treated both Upper East Side elites and Jews from my city's own shtetls: Crown Heights, Borough Park, and Williamsburg, Brooklyn. Because he was a polyglot, he was sought after by the city's ultraorthodox, the Hasidic Jews who live in those insular enclaves where Yiddish is the vernacular, to treat them in the languages they spoke. He saw them largely for free, palming the poorest of them subway fare to flee their ghettos of Brooklyn for his office down the street from our apartment at 903 Park Avenue. Under their head-coverings, fur *shtreimels* for the men and *sheitels,* wigs worn for modesty to hide women's own hair, was strife—just like any other New Yorker. Often it was underscored and exacerbated by the repression demanded by extreme religious observance.

For them, Dad offered his favorite diagnoses-by-catch-phrase. One was "A sense of mastery." What we were all looking for, Dad said, was the feeling that we achieve only by mastering something, and he exhorted his patients and me to take full command of our lives. Those endemically human feelings of being lost, rudderless, unmoored, Dad believed, are the result of not giving oneself permission to seek out mastery. Fully living was not just making one's place in the world, he said, but mastering it.

"Belonging and not belonging" was another favorite—a paradox that he believed was the root of so many of his patients' suffering. For the largely impoverished Hasids, belonging and not belonging was the struggle of remaining pious anachronisms in a modernizing society. In his own daughter, belonging and not belonging was inescapable as well. It was why my mother had torn up the wiring on the grandiose toe buzzer beneath the dining room table in the Park Avenue apartment. She was *in* the apartment but not *of* it, her actions insisted.

I experienced it as an essential tremor of unworthiness, an electric current that pulsed one word like neon behind my eyes: *outsider*.

Dad's highest compliment was "authenticity" (though when he said it with his muddled accent, it was "auTENTicity"). He sprinkled it in conversations as a way to praise worthy art and laudable people—those who cracked the veneer of society's expectation and let their true selves shine through. When his Orthodox clients understood themselves and under his guidance let themselves feel feelings verboten in their clan but necessary for self-realization, they were at last authentic. Authentic, they at last had a chance at happiness, he believed. But who was my *autentic* self?

Dad was the ultimate at belonging and not belonging. That's because his early existence, in a way, was predicated on inauthenticity: to survive the Nazis encroaching on Lvov, Poland, where he was born, he obtained myriad false identities. The first pseudonym was inked on a forged baptismal certificate made by an artist named Ludwig Selig. Thanks to Ludwig's expert penmanship, dad became a fictional Catholic Pole named Julian Heybowicz. The new identity denied his Judaism and hid his faith, but it saved his life. For his efforts at saving my father and others, Ludwig was murdered.

Dad's own name, Yehuda Nir, was not really his. My father was born Julius Gruenfeld, a name with deep Germanic roots. In 1952, in the aftermath of World War II when he was twenty-two, he changed it, so, he would say, archly, "no one will see my future success and attribute it to a German." Nir, our new surname, means a plowed field in Hebrew—roughly the "green field" that Gruenfeld means in German.

Dad's first name became Yehuda: It means "Jew." Authentically himself at last.

The chance to be authentically a horse person stirred me with excitement when he mentioned our roots. I couldn't believe my family in fact belonged to the world where all I felt, as I trotted around each riding ring posting, that is, rising and falling with each step, was the thump-thump-thump of not belonging. I sat up on my heels and pressed him for more.

"It was World War I," I recall my father said. "You had a cousin. He was crafty. He was a pacifist. He hated the war—the troops trampling through the farmland, crushing his crops, enlisting his sons. But of course, you had to pretend you loved it." He paused to pull from his cigar. "Actually, he loved the war a little. Because when they were slaughtering one another, maybe it was a distraction from harassing the Jews—you know, they hated us even then," he said with a laugh. "But your shrewd cousin of course, pretended he was an autentic nationalist.

"For a while, the armies camped in the woods near his farm," Dad continued. "They drank and left a trail of gaunt and lame horses behind them. Each poor, limping, half-dead *koń* they would abandon like a dog in a junkyard. Such a *balagan*," he said, Yiddish for a total mess. At night, Dad explained, the cousin would go into the forest of what would later become Poland. Under the cover of darkness, he would collect the used-up horses the various cavalries traipsing by had let loose to die in the woods. Broken horses burned out from the endless marches, exhausted horses returned from the front.

"And he saved them? So he rehabilitated them?" I asked. This long-dead, mysterious pacifist cousin was sounding more and more like my new best friend.

My father puffed the cigar. As close as my father and I got as he aged, he was lordly when I was a small child, softening to the American style of interplay and engagement with one's children only when I was long past childhood. He had been the chief of child psychiatry at one of the city's foremost cancer hospitals, but was better at diagnosing and healing the minds of sick kids than playing tag with us. In his balky efforts to play American dad to one of my brothers, he once hired one of our doormen to hit baseballs with his son. I think Dad thought it was the best he could do. After all, the memoir he wrote of his experience hiding from the Nazis from ages nine to fifteen is titled *The Lost Childhood*, and when your childhood is stolen by war, you never learn how to play catch. But his awkward efforts only made his children feel more distant, stranger.

Dad continued with the story as if he had not heard me. "In

the morning, he'd send a runner to the army's tents with word that there was a fresh crop of horses in his paddocks: Robust warhorses ready for battle!"

Soon the troops came horse shopping. "Your cousin would post a lookout at the far fields of his farm, and as the cavalry approached, he suddenly made the half-dead horses come back alive," my father continued. I waited breathlessly to be regaled with a tale of familial equestrian prowess. I was eager to run back to the barn and tell the little girls whose living room walls featured daguerreotypes of their grandmothers astride fox hunters and oil paintings of great-grandfathers mounted in military dress that I had a cousin who was a horse-whispering Polack. Dad shouted into the living room: "Pepper in the ass!"

I was aghast. The cousin's trick was a dash of capsaicin—the stuff in chiles that makes you sweat—under the tail and the half-dead animals would buck and rear. They were wild in his pen at the irritation despite their sapped strength. These raging, seemingly battle-ready mounts impressed the soldiers, and they snapped them up. In reality, they merely bought back their own decrepit horses. My cousin relished the sales not just for the cash he earned, so the story goes, but for the subversiveness. By furnishing the armies he hated with animal equipment that fell to pieces the moment the fire under their tails burned out, my cousin waged war on war.

My father adored telling the story throughout his life. Whenever anyone asked about how his daughter became so infatuated with these animals, he'd laugh, "It's in our blood! And under the tail!" The story's theme of subversion delighted him most of all. In telling it, Dad turned the trope of the greedy Jew on its head. This long-lost cousin was a profiteer, of course, but in my father's retelling, Shylock was the hero. My father reclaimed the cunning of the Jews—the terms of disparagement that have been lobbed at our people since time immemorial—as a mighty weapon. Jewish guile was weaponized in both the fable of the pepper and my father's own survival during the Holocaust.

"It felt great to outwit 11 million Germans trying to murder a nine-year-old boy," goes a quote from my father used in his *New York Times* obituary when asked, once, about his life story. "I present it as a psychological victory."

Dad's myth of the pepper omitted names, dates, and places, and I never got the chance to plumb Dad for the meat of it—truly a journalistic failing. But recently I decided to find out if the story was in fact true.

It was August, a slow, sticky Sunday shift at the *Times* headquarters, where my beat then was what is known as a rewrite reporter. It is a term from yesteryear, when reporters called into the marble offices of their broadsheet with frantic details from the courthouse or the crime scene. In my imagination, those reporters all spoke solely in that nasal patois of old-timey steeplechase announcers. In the newsroom would be a sweating rewrite man, hunched before, say, a Sholes & Glidden typewriter, taking dictation and banging the story into newsprint.

I use the word *man* intentionally, because for much of the history of newspapers, writing was precluded from women. There is a giant photomural of the 1950s newsroom, which is exclusively white and all male, poring over their typewriters. Most recently, it decorated the corner office of a black, female mentor of mine. For a while, she considered replacing it, until we concluded it was more satisfying to have them on her wall, watching her run the show.

In the modern era *rewrite* means feeding the gaping maw of the Internet with almost instant information on whatever news is breaking. On slower days, I would tackle things like the newsworthy deaths of people who didn't deserve what they got. It is an agonizing duty: calling people on their worst day and asking them to help me weave the filaments of a life lost. When they hang up on me and curse at my work as a vulture gorging on the carrion of their suffering, I agree completely. When they take a tremulous breath and then unspool the story of how their loved one lived—a knack for baking butter cake, the funny burble in her laugh, a favorite sweatshirt he never took off—and thank me for the grace of giving a life to the death, I agree completely too.

On particularly hard calls, I pull up a picture of a random galloping horse on my screen while they speak, to hold part of me in that space that takes everything away.

That Sunday, no news was breaking. Beneath the light of the flickering Times Square billboard out my office window, I was beset with a sudden curiosity. It had never occurred to me until then to ask if the spicy practice my father described was in fact real. I looked over my shoulder to make sure no colleagues were there and tapped out the three words into my browser.

"Pepper." "Horse." "Anus."

(As a reporter in this era, at a time when the presidential race itself was hacked, "Email like the Russians are watching" has become something of a personal credo. I enjoy imagining some Vlad or Sven snooping into my search history and finding my spiral into the world of horse butts.)

I found illumination in an entry in a two-century-old book, the *Classical Dictionary of the Vulgar Tongue*. It was published in 1785, and written by a lexicographer, Francis Grose, an Englishman who gathered the lingua franca of seamen and prostitutes on midnight walks in London's dockyards and dark alleys. From it, he created an encyclopedia of slang. With his assistant (the delightfully named Tom Cocking), Grose jotted down plenty of normal things, such as some of the first written references to tidy phrases like "birds of a feather," And plenty of really weird stuff too—just what I was looking for, in fact.

"To Feague A Horse: To put ginger up a horse's fundament, and formerly, as it is said, a live eel, to make him lively and carry his tail well; it is said, a forfeit is incurred by any horse-dealer's servant, who shall shew a horse without first feaguing him." Grose adds this helpful addendum: "*Feague* is used, figuratively, for encouraging or spiriting one up."

But it is a real practice to this day. I know this because the next listing down in my Google search was something called "Plummers Tail-Set Ginger Salve." Curiosity popped it into my online shopping cart—$4.50. The little glass tub arrived a few days later. Slapped on its label is: "DIRECTIONS: Apply around horse's rectum 10 to 15 minutes before entering show ring." Feaguing is today called gingering. It is a dubious practice cultivated primarily in one of the riding disciplines with which I am most uncomfortable.

Tennessee Walkers are a unique American breed: high-stepping animals with necks that arch almost inversely, rising up high before the rider. Astride, a rider's eyes are almost level with the animal's crown. The tail floats behind the horse, carried high and fluttering away from the body like a pennant in the wind. And they prance. These horses are considered "gaited," because they are selectively bred over time for unique movements over the various paces. They number among a handful of other gaited breeds like the Peruvian Paso Fino, with its *paso largo*, a butter-smooth quickstep in double time, or the Icelandic Horse's rhythmic *tolt*, a gait that gives it moves like Travolta.

This discipline of gaited horses is all about the strut. In an ideal universe, a Tennessee Walker whips its hooves skyward and flutters its tail aloft because it is full of spirit and bred to stride with a militaristic march. But too often the flamboyance of their paces is the product of a litany of devices and practices that to me have sullied the discipline.

There is soring, the use of caustic chemicals and chains on the ankles of horses whose aplomb is desired, applying stinging and burning agents to make them prance ever higher. It was targeted in 1970 by federal law, but a dearth of animal welfare inspectors and funding to carry out the mandate means it still happens. To create the "big lick," as the exaggerated prance of the Tennessee Walker is called, the horses are often fitted with weighted wedges, known as "stacks," strapped to the hooves. Like a person striding through snow, or wading through the tide, stacks weigh down the legs, thus demanding an exaggerated thrust into the air from each stride. Stacks also can hide soles cut to the quick to make the horse hot-step in its effort to avoid the pain.

Even some of the permissible practices for the discipline fill me with horror. "Nicking" is severing tendons in the tail, so that with corset-like training it will be carried higher, with more panache. An 1838 treatise on animal husbandry calls tails cut in this way "mutilated," and even back then wrote, "It is a surgical operation, but no respectable veterinarian would recommend it." And yet it continues. At a barn I rode at for a time in California, grooms would tether one Walker's head to the ceiling beams. It would remain tied for hours, its back arched in readiness for when its master came to ride. It was a national champion.

Who am I to cast judgment? I often have battled with the ethics of this sport I love. Is what I do, burdening an animal with my body, cajoling it with the threat of whip and spur, making it work at all, just a variation on the theme?

For a while, my father had no memories of his life between the ages of nine and fifteen, his childhood in Poland in the maw of World War II, from 1939 to 1945.

Then one evening when he was nearly sixty, the war long past, his life in America well under way, he was invited by an acquaintance to a fancy dinner in Connecticut with his young wife, my mother. The two doctors were invited so that, the host crassly explained, "the other Jews at the party don't feel out of place." Seated beside one another, my father chatted to a man who had a funny, lilting accent, not quite British. With his penchant for languages, my father began to grill the man, as he did whenever he heard the music of a person's voice, trying to draw out that person's story.

In the posh living room in Westport, the man, Jack Mausner, explained he had been raised in the United Kingdom, fleeing Poland after the war. His Queen's English was lilted with Polish that would never go away, even though he had never returned to his hometown—Lvov.

"Lvov?" my father said, hearing the name of his own birth-place. His breath stopped. He looked deep into Mausner's eyes. "Kuba?" he said, using the man's childhood nickname. "From Hebrew school? Kuba, is that you?"

On the way home, my father paced the deck of the ferry from Connecticut across the Long Island Sound en route to our house in East Hampton. Five decades and over four thousand miles from the small Polish city where he was born, he had dis-covered Kuba, Jack Mausner, one of the only other survivors of the Hebrew school he had attended as a boy.

"Murdered," Dad was always sure to say, never "killed" in the Holocaust, never "died." "Soldiers are killed," Dad would tell me and anyone who would listen, including every newspa-per he came across where the wrong word was used. Whenever he read it, he would fire off a scalding letter to the editor, at least once a week until he died at age eighty-four. My mom keeps a collection of them in a binder wryly entitled, "Letters from an Angry Jew."

There in Connecticut was a playmate resurrected from the ashes of those halcyon days before the Germans swept in and murdered my grandfather. The days before my nine-year-old fa-ther, my teenaged aunt, and their mother fled their cushy bour-geois life into years of hiding. They hid in attics and in plain sight: they survived with forged baptismal certificates and fake identities as Catholic Poles.

There in Connecticut came years of stories my father had never told, even to himself. Perfectly crystalline, still there.

That night he didn't sleep—and neither did Grandma Frieda. He woke her up, and in her paisley dressing gown, she followed him to the basement of the beach house and dusted off her old typewriter and her secretarial training. My father began that night to dictate to Grandma what would become *The Lost Childhood*, his acclaimed memoir of how he survived Adolf Hitler. It came in an onrushing torrent of memories my grandmother sopped up as scribe, click-clacking his once-erased history into the night.

I need to stop here and say what I once said to my father: that by comparison to him, I don't feel like a real person. It seemed that only my father could be, forged in the fire of war, made canny by needs as atomistic as survival, and that everyone else with nice lives, and nonmortal traumas, was necessarily a soft-bellied simulacrum.

"Never envy me my wounds," my father said harshly, setting me straight. "I envy you are spared your own."

My life with horses has induced at times deep survivor's guilt, as can, frankly, things like general well-being, a roof over my head, a good job, and even a nice sunny day, when you have a father who was a victim of atrocities and thus feel no right to any ease of your own. Even writing intermingled paragraphs about ponies and Nazis feels off. But they are all the threads of my story that, braided together, form my life. And I must remind myself that my father himself never compared battle wounds and found others' wanting. As a psychiatrist, he specialized in treating posttraumatic stress disorder, validating the scars of others, whatever their source.

And to him, as to me, horses were freedom. On April 28, 1945, my father was fifteen. His tattered family had been impressed into hard labor on a German farm by Nazi-sympathizing farmers who believed they were harboring Polish refugees, not Jews. Suddenly, on a ridge, were Russian soldiers. The war was over.

My father, his mother, and his sister rode away from the farm to freedom in a *britzka*, a four-wheeled carriage, pulled by "the only horse left in the stable, a two-year-old mare that hadn't been trained yet to pull a carriage, and therefore had been left behind by the Russians," he wrote in *The Lost Childhood*. "We were all happy except the horse, who was having a hard time adjusting to her first job."

And so throughout his life, the scent of horses conjured freedom, full of relief, full of peace. He'd sigh deeply into my hair when he hugged me after a long day of mucking stalls and grooming horses brought the barn home with me. And on Fifty-Ninth Street and Fifth Avenue, among the Central Park carriage horses and the scent of his remembered liberation. When he drove to the stables to pick me up from doing chores or riding horses, he'd roll down the window and close his eyes.

THE DAWN HORSE

I grew up outsourced to the nannies who raised me and to the city itself, a place that kept my insatiable curiosity slaked, if not ever fully satisfied. As I was always my best self around horses and there were few to be found on Eighty-Seventh Street and Park Avenue, my nanny, Beverly, would schlep me to the next best thing: the American Museum of Natural History, where I'd spend long hours inhaling the history of the horse.

There, the white bones of horses gallop in a cavalcade across a gallery on the fourth floor, ancestors trailing ancestors to the front of the herd, where the modern horse leads the charge, his skeleton mid-bolt. When I'd find the horse bones, I'd stare and stare, ducking away only to gobble a paper boat of fries and lust after the dinosaur-shaped chicken nuggets that, sadly, weren't kosher, in the basement cafeteria. Crumbs falling, I'd skip stairs back up to the exhibition, marveling at the skeletons, imagining the 55-million-year-old animals enrobed in flesh and fur and me on their backs. I'd always arrive hoping archaeologists had dug up a fossilized unicorn.

The horses helped Beverly keep me busy, and she needed to.

In our plush apartment across the park from the museum, no one was usually at home but us. In New York City, my father's linguistic skill that sheer survival necessitated, coupled with his personal understanding of trauma, made him famous. He was not only the go-to therapist for the city's sprawling, ultra-Orthodox Jewish community, but an associate professor at Cornell University Medical College. When I was little, he became chief of child psychiatry at Memorial Sloan Kettering Cancer Center.

Dad was busy.

My mother is a psychologist, so outspoken she makes Dorothy Parker look demure. She's the loudest person in the loudest room you've ever been in, always; but she's also a deep listener, and her advice is so sage that it's almost law. She is blonde and green-eyed and, in the 1980s of my babyhood, the perfect package for the newly emerging phenomenon of glamorous TV therapist, prefiguring Dr. Phil by decades. She and my father co-wrote books on how lovelorn singletons might find spouses, and then sequels on how to have a peaceful marriage (never mind that most of my parents' conversations were conducted in the key of an enraged bellow). My mother hawked the advice on television, sharing the stage and bad perms weekly with that other nascent '80s creature: the daytime talk show host. She was one of the first guests on the first year of Oprah's first talk show.

I mostly saw my mother on TV: in the kitchen, above the microwave oven. There, I sat to eat Empire Kosher chicken nuggets Beverly defrosted after we returned from our sojourns at the Natural History Museum, and we watched Mom chat with Oprah, Geraldo Rivera, and Phil Donahue on the screen.

On one of her first appearances in 1986, Donahue brought her on air to quiz her about the "problem" of the newly emerging class of "career women" who were not getting married off. Under her bob, you can almost see Mom's shoulder pads bristle.

"I have seen a revolution no doubt, and the people who gained most in the revolution have been women. But you know what? It's made things much more complicated. So it is quite true I am seeing a lot of relationship problems—but when women walk in with those problems, they walk in more confident, more successful, more powerful, and better off in many ways."

Growing incensed, Mom half-stands in her chair. The audience stares. "And it would be a disservice to this revolution to suggest that this complicated problem is really something that is a kind of punishment exacted on us for, really, getting a little 'uppity!'"

The studio audience burst into applause.

Mom was busy.

There are tens of thousands of horses on the private fifth floor of the American Museum of Natural History. Well, to be more precise, there are tens of thousands of what horses once were. Housed there are iterations of their ancestors galloping backward through 55 million years of time.

I didn't know about the herds of hidden horses in the upper recesses of the museum when years ago I hung out in the marble public gallery eating leftover fries that Beverly hid in her pocket and snuck to me one by one as I stood rapt. Beverly Moore was born in Jamaica, and she's why, when I was little, I spoke with a slight Kingston accent whenever I was angry. I spent almost every waking hour with her, patois our secret language we spoke so Mom couldn't understand us over our meals of island staples like saltfish and ackee in the kitchen on Park Avenue.

Beverly had six children of her own, starting when she was seventeen years old. Back home in Jamaica, she was a "slippers maker" as she called it, a shoemaker who specialized in elaborate sandals. As she sat on her porch there one afternoon, sewing pieces of leather together, she looked up to see four of her children walking down the path to the house. They were carrying her slipper supplies from the shop where she'd sent them to pick up the goods—leather thongs, new buttons, and buckles.

Bev dropped what she was sewing. Like their mother whose childhood flamed out even as it started, her adulthood accelerated far too fast with the jet fuel of teenage pregnancy, Beverly realized with a start that her young children were already working. She saw on the path they walked toward her on that day an inexorable route to repeating her life. It was one she wouldn't trade—she had Maxine, Melisa, Milton, Marlon, Marcia, and Monique, her treasures, to show for it—but also wouldn't wish on any of her children.

That month, she got on a plane to New York City. As she cleaned houses and obtained a green card, she brought each of her six children, one by one, to join her. In June 1990, Bev met my mother, who hired her to care for me, and the two began a symbiotic partnership best summed up by the phrase they stand together and repeat at the head of the table every Passover. There, over a meal of Beverly's matzoh balls, my mother reminds everyone that in the Exodus story, the person who saved baby Moses from the Nile River was Bitya, the daughter of the Egyptian pharoah—an outsider who cared for a Jewish child as her own. "The heights we reach are measured only by the shoulders we stand on," my mom and Beverly recite side by side.

On a trip to Park East Kosher Butchers to pick up brisket for Passover one year when I was about twelve, Beverly met a hunky butcher named George Moore who kept sliding her free cold cuts over the glass countertop. They were married, and she became an American citizen.

She was an immigrant like my father, but by the time I knew him, he had long ago achieved his American dream and was living it with relish. As I walked beside Beverly while we wended our way through the city together, we spoke about Jamaica or about horses, but what we really were discussing were our dreams. Beverly was mid-battle for her American dream, and witnessing her fight for and win it was the first time I felt the importance of the untold stories of everyday heroism that shape our lives. Telling those stories is now what I do for a living. I wanted to be like her. I still do.

Beverly provided the rung that allowed my mother to climb to the top. And my mother was her ladder in return, but it was Beverly's children who ascended. Today, all six have achieved secondary degrees—seven, because I count myself as Beverly's child.

On a rain-drenched afternoon recently, I stood inside the museum where Bev and I had spent so many hours. I wended my way among the bodies and bones, hooves and hairs, up to the hidden fifth-floor collection. The existence of this upstairs warehouse of thousands of horses had been unknown to me. It is reserved for scientists, not horse-mad little girls and their nannies.

I had asked the curators to grant access to the labyrinthine upper stories of the museum because I wanted to find the first horse of all time. My guide into the past was Dr. Ross D. MacPhee, the curator of the Department of Mammalogy, who is as passionate about lemurs, his specialty, as I am about horses. Dr. MacPhee studies the thousands of horses that lie in state in its archives. He was quiet and serious, his chin dusted with white hairs aspiring to a beard under half-moon wire spectacles. He greeted me in the doorway to his office, which was flanked by two elephant tusks that stood on end like ivory parentheses, and beckoned me to step through the door. In the recesses of the museum's scientific collection are scads of fossils—ancient insects in amber, Mongolian dinosaur eggs—but only one animal has such an extensive collection devoted to it, in fact the largest anywhere in the world: the horse.

The silent rooms were filled with padlocked metal chests— "cans" in museum parlance. Each was packed with slim wooden trays in which preserved eye sockets, hock joints, and cannon bones rested precisely on sheets of white paper, a library of bones. Beside each item was a yellowing handwritten oak-tag card scrawled in ink, paper tombstones for the ancient equine ancestors each cabinet contained.

Here lies Hyracotherium. Rest in pieces, Palaeotherium.

Dr. MacPhee pulled out a drawer and popped a 55-million-year-old skull into my hand. This was from the animal that pre-figured horses, Eohippus, also called the Dawn Horse (dawn, as in, the dawn of time). I stared at its polished teeth, gleaming from the skull. The molars were black; the pigments of the rock in which it survived history had leached into the dentin and enamel. It was as if the proto-horse was of the mineral earth itself, the way Adam and Eve were scooped out of dust.

Horses predate Eden. They predate humans. The modern horse has been around for 2 million years; we are only just about 350,000 years old. And the antecedents to the equine, like the one I palmed that afternoon, can be 55 million years old.

I turned the little head over in my hand. How did *this* grow from cat-sized creature to the modern—rideable—giant? "Grass," Dr. MacPhee said. "It's all about grass."

Horse ancestors are older than grass, which sprouted only 15 million years ago. Before then, animals subsisted on fruit and leaves they tore. When grass took hold, the animals began to evolve. Tough grass requires a cavernous vat in which to be fermented by the body and broken down into a nutritious slurry, Dr. MacPhee explained. As grass spread throughout the world, horses grew in size as their bodies were forced to become ever-larger tubs to process the new food.

(I drove out to the barn in New Jersey where I kept my horse, Bravo, later that afternoon. The rain clouds had fled, the paddocks damp still. I bellied down in the grass beside his muzzle. All his vastness owed to millimeters of green? Those tiny unthinking blades gave horses height and breadth enough to bear human weight. I ran the blades through my fingertips, and thanked the first bold seeds that dusted an ancient universe for the horses they bore.)

Horses are some of my first memories. And they are some of the first memories of humanity as a whole. Over thirty-two thousand years ago they were painted onto the walls of a tide-washed nook in southern France known as the Chauvet Cave. There, blond buckskins, red roans, and even spotty appaloosas, colors you still see today in any barn, tramp across the limestone. Their manes are broom-bristle-spiked like those of zebras, and they arch delicate necks to stretch their muzzles across the stone wall, side by side, still in soft charcoal that sears through time. These are the earliest artistic depictions by humans. And they were of horses.

About ten thousand years ago in North America, horses went totally extinct. While we associate horses with America's soul, they're actually a recent import, reintroduced here in the belly of a Spanish galleon by conquistadors after 1492. Once, as the little skull in my hand that day at the museum and the thousands in the cans around me attested, horses had the run of America. But something went mortally wrong for American horses about 8000 BCE. Across the continent, they were suddenly gone.

When I first heard that fact about a decade ago, I could not believe it. I thought of the Plains Indians on their painted ponies. How could it be they had never ridden—no, had never seen—horses until the sixteenth century? How was that possible when Indian equestrian prowess is inked on buffalo hide and indelibly in the American psyche? How could something so quintessentially American as a horse be so recent an import?

More than that: What happened to the horses?

Dr. MacPhee? "No one really knows," he said flatly as we closed up drawers of horse chunks, and I grew quietly enraged at the scientific community. *No one really knows?* We can put a human on the moon, and yet millions of horses just went poof ten thousand years ago and no one can figure out why? "There are theories," Dr. MacPhee piped up, perhaps sensing my quiet unraveling as we wended our way through the narrow aisles of the fossil catalog and back to his office. "A mini ice age," he said, listing theories, "or a plague."

As we reached the elephant tusks that bracketed his office

door, he whirled and suddenly jabbed his finger in the air at me with another theory: "Or YOU."

Humans crossed the Bering land bridge and spread into the American continent about sixteen thousand years ago, Dr. MacPhee explained. We were predators. I reeled a little there between the ivory. Were horses hunted to death? Is the Chauvet Cave just a pictographic meat locker?

And there are still more mysteries in the horse's story. Just as scientists still search for the so-called missing link that led to *Homo sapiens*, the fossil record has not yet disclosed all the creatures that led us to the modern horse.

I scribbled in my notebook as Dr. MacPhee explained this all to me, using a pile of roughly hewn wooden boxes set at odd angles beside me as a desk as I wrote. They were stacked haphazardly, in stark contrast to the precise cans of carefully labeled hooves and heads. The boxes contain some of the archaeological finds of a socialite and explorer named Childs Frick, an amateur paleontologist who traveled the American West in the mid-1900s. Frick and his team scooped up more than two hundred thousand fossils, horses and proto-horses, lodged deep in the country's soil. I was stunned as Dr. MacPhee revealed why the boxes were plopped there: a century after Frick hauled the horses and other mammals home, no one has yet gone through all his finds.

Some of the crates are scrawled in black ink with the word *horse*. They were still nailed shut.

"So in these boxes, here in the middle of Manhattan, just piled here beside us, could be the equine missing link?" I gasped, pulling my notebook off the heap of old crates as if it were a hot stove.

"Or maybe," Dr. MacPhee said, "it's a unicorn."

BREYERS

My kingdom for a horse. I would have given anything for a horse of my own, a Cutie or a Guernsey who would hold me in that same spot in their heart in which I held them. I wanted to be responsible for a horse the way that Grandma and Bev were for me and have something love me as hard as I loved. I wanted an anchor covered in fur.

But horses cost a fortune. There's a joke: "How do you end up with a million dollars in horses? Start with ten million dollars." It's not necessarily the initial investment in the animal (though there are show horses that cost a quarter of a million dollars and up, as well as rescues that cost nothing); it's the upkeep: even the cheapest horse in the world eats about the same amount in sweetfeed and hay as a priceless Triple Crown winner.

Hooves grow like fingernails and must be trimmed or shod every few weeks; in my neck of the woods, that means a couple of hundred dollars every time. Stabling and training, particularly in the posh beach town my parents had barely eked their way into, are often exorbitant. And horses can live thirty years or more, each year of increasing decrepitude requiring more and more costly vet care.

"Men work more for horses than horses work for men. Such is the conclusion we reach after reading the latest data on horse chores and horse-feed and horse costs," Herbert N. Casson wrote in *Munsey's Magazine*—back in 1912.

With little backyard on our eighth of an acre by the sea and my parents *unconscionable* refusal to stable Cutie in the bathtub in our New York City apartment, I knew even as a little girl that horses were beyond the reach of my striving family.

Fortunately plastic horses then cost twenty-six dollars.

At a 1 to 9 scale, the perfect plastic replicas, called Breyer model horses, were my solace and fixation. Every birthday as a little girl, all who knew me trucked to F.A.O. Schwarz or Miller's tack shop on East Twenty-Fourth Street to buy me one of these realistic equine sculptures, to my rapturous joy. Whenever the school bus–yellow of the classic Breyer box peeped from the wrapping paper, all my unsatiated equine ardor poured into the plastic creature. The horseless years of my young life, my imagination was my solace—whether my Cabbage Patch horse was towing me in my Radio Flyer wagon across puddles in the yard or I was making my cocker spaniel Electra jump overturned chairs in the living room. For me, Breyers came to life.

Some part of me swirled a darker message into the playthings: Breyers were the horses to which a girl like me, I believed, was permitted access. The world where you had a real horse of your own seemed reserved for other girls. Theirs was a universe of flaxen Ashleys and Charlottes like my Brearley classmates. The kind of girls I imagined routinely open snowy barn doors to see a pony wearing a drooping red bow, shyly peering out from under a shag of forelock. Girls with accentless fathers who celebrate Christmas and never have to bring damp, buttered matzo in plastic wrap to school for lunch.

Since Breyers were as close as I ever thought I would get to the real thing, I reveled in them like they were. Of course, they were clattery cellulose acetate that would never snuffle my hand, so the high was temporary, in sharp contrast to the high of a real horse, which is indelible.

As I grew up, I spent all my babysitting money on Breyers and eventually had a stable of sixty-five. They're still in my mother's basement. I'm not alone: today over a million Breyer model horses are sold each year.

One afternoon when I was thirteen, I walked into a new tack shop that had just opened in East Hampton town. It was across from East Hampton Bowl, the circa 1960 bowling alley where I used to spend rained-out afternoons rolling gutter balls with Grandma. Now that I was a teenager, I had been hitting up the shiny pinewood lanes on Friday evenings for new reasons: to ogle boys, whom I had recently discovered were interesting for some reason I could not yet figure out.

The store was in a shopping esplanade fittingly called the Red Horse Plaza, featuring a bucking vermillion horse on a big sign out on Route 27. In the pasture just behind the white-brick shops in the complex, a pair of mammoth Clydesdales had lived for years. They were village fixtures, carefree in a field of tall grass that obscured the bell bottoms of their white-fluffed legs until they mowed it all down with their teeth. They munched oblivious to the parade of Mercedes, Jaguars, and BMWs speeding by their paradise along the thin highway that zips through all the towns of the Hamptons. They lived incognizant of the wealth all around them, as horses thankfully always do.

I pushed open the swinging door of Red Horse Equestrian with one thing on my mind: horses. Plastic ones.

The tack shops I had known and haunted were homey, pragmatic places like the Tack Trunk in nearby Amagansett. There, an old pony trainer of mine, Leigh Keyes, presided over buckets full of things horses actually need, like brushes and lineaments and unctures and tubes of deworming paste. Behind the register at the Tack Trunk was a pushpin wall of photographs of her old students, little me included.

Red Horse Equestrian had artisanal everything, and everything was useless. There was cashmere gear unfit for horse slobber, and riding boots that cost so much that when I told my mom the price, it sent her into apoplexies of Yiddish. Getting shoes that expensive anywhere near manure, as they were intended, was a *shanda*, she said, a scandal!

Within moments of walking in the door of Red Horse, I learned that the shop owner knew roughly as much about horses as does my mother. To this day, she calls a horse's halter its "collar," as if it's a spaniel. (Occasionally she takes the slip-up further, asking me if I had a good ride on my dog at the barn.)

He knelt in a corner beside a girl of roughly eight, attempting clumsily to buckle a strap of leather around her calf. Backward. The strap is called a garter, a slim circlet that goes around the lower calf. It is an item unique to the competition uniform of the very youngest English-style riders. Garters are a relic, like much of equestrian sport, from another era. Once upon a time, garters, like horses in general, were useful. Prior to the advent of high-tech stretchy fibers, garters held the loose fabric of riding britches down to the leg. They prevented bunching between the saddle and the knee, which inflicted abysmal rubs.

In the age of lycra and nylon fibers, garters are as vestigial as an appendix. But enter a short stirrup competition ring without them, and the most perfect cherub of a rider will get points docked for sloppy turnout. And here was a man selling a newbie competitor her performance gear, the minutiae that separates, in the eyes of a judge, the disciplined horsewoman from the unscrupulous, and—gasp—teaching her to put it on upside down!

It was all too much for this thirteen-year-old equestrian fanatic to handle. "Hi, I'm sorry to interrupt," I said, "but you're doing it wrong." They were inauspicious words to the man who was within minutes to become my first real boss, but they did the trick. He stepped aside and let me suit the little girl up.

Once I buckled her into the garters, I took her by the hand from rack to rack, pulling out each item of the show uniform she would need. Into her doting mother's arms went a tiny woolen suit jacket in subtle herringbone and a velvet helmet I measured and fitted snugly to her head. In a corner rack, we picked out a pale pink cotton blouse with a detachable priest-like collar called a ratcatcher. Another antiquated part of the English competition uniform, it derives its name from the uniform worn by those employed in Victorian-era pest control, so the story goes. Those were men with terriers trained to go to ground for vermin, who for some reason appear to have dressed rather nattily. When we got to the register with the pile of gear, her mother pulled out her credit card. She was a granddaughter of John F. Kennedy. The mother turned to me and asked: "Now, where can we get her a pony?"

I worked at Red Horse Equestrian from that day onward that summer, a year too young to be legally employed (I'm still not sure how the records were fudged). But I need not have worried about any tax dodge: I essentially wanted to be paid in Breyers.

What began as the Breyer Molding Company in Chicago didn't set out to make toys for which horseless horse girls went gaga. The company crafted plastic doodads and useful thingamajigs like knobs and eyeglass parts. In the late 1940s, Mastercrafters, a clockmaker, contacted Breyer and proposed it whip up a decorative equine clock-topper. Breyer ended up running with Western Horse #57 as a stand-alone objet d'art. It was wildly popular and proved durable as a toy; today Reeves International, the New Jersey–based company that acquired it, crafts hundreds of different models in a process that takes eighteen months from concept to Connemara.

Along the way, something interesting happened: people started competing in contests—you read that right—with their plastic model horses. So did I. When haranguing my cocker spaniel to jump over lawn furniture in East Hampton failed to fill my horseless itch, I would slip down to the beach with a half-dozen Breyers tucked under my arms. On the rocky revetment that kept our house from falling into the sea when hurricanes whipped through, I posed the models and imagined them as wild Arabians in the desert. With my mother's Polaroid camera, I'd strive to capture the most picturesque shot, careful not to get sand in the lens.

The purpose was to enter my plastic horses in competition

in a "photo show." *Just About Horses*, a little magazine put out by the Breyer brand, provided mailing addresses of so-called shows in the classifieds. A photo show consisted essentially of a stranger who decided to judge photographs that random people mailed to them of Breyer horses. I'd check the mail every day after I sent my pictures in, waiting for the ribbons they'd post back to the winners.

But my childish Polaroids were nothing compared to the lengths gone to by serious Breyer enthusiasts who cart their acetate herds across the country for live (so-to-speak) model horse shows. "Showing" the models has been around since the 1970s and wasn't something that the company had planned. "The horse showing, that was fan-driven—we didn't think that up," said Stephanie Macejko, a vice president of Reeves, the parent brand.

"It's a tough one to explain," Macejko said as we wandered around Breyer headquarters in Pequannock, New Jersey, about thirty miles northwest of New York City, on an afternoon not too long ago. It's a nondescript office building, improved by the silhouettes of horses galloping toward the glass double doors. "For our fans, it isn't just a toy," she said. "It becomes a real horse to them."

I followed her past dioramas of acetate horses splashing through resin streams and vaulting jumps overgrown with minuscule paper flowers. She unlocked a door to a library of shelves that fairly heaved with model horses. The horses were spangled and in every hue in the natural world from Appaloosa to zebra, and some not; one glowed in the dark (there's a Halloween special edition named "Night Mare").

I pulled out my phone and began taking pictures of this toy

shop from my childhood's wildest imaginings. Macejko hastily closed the velvet curtain, blocking me. I had not realized that behind it was an aisle of horse models in preproduction, not yet released. "Top secret," she whispered.

The horses are accurate to a point. Sometimes a pure replica doesn't in fact capture the animal, Rick Rekedal, Breyer's executive vice president, told me as I took a seat in a conference room next to a shelf piled with thumb-sized unicorns. What the Breyer artists create, he said, "is not just the exact sculpture and scale of that horse, but also how that horse makes you feel."

Not everyone agrees with the artistic liberties. Many Breyers are modeled after famous animals. Once, dissatisfied with the hue of a reddish bay on early versions of the champion Standardbred, Niatross, his handler showed up in the parking lot in Pequannock with the stallion himself to show the artists his true color.

Breyer didn't come up with horse shows for its models, but the company has fostered the hobby and galloped with it. Since 1990 it has thrown BreyerFest, the lollapalooza of the model horse world, attracting giddy collectors from across the globe annually to Kentucky Horse Park in Lexington. They come armed with carloads of plastic ponies to show off and sell. In recent years interest in the pastime has risen, Breyer executives told me, driven by social media: instead of mailing in Polaroids, hobbyists now compete by posting vignettes to Instagram. The numbers of attendees at BreyerFest have exploded; more than thirty thousand model horse collectors converged on Lexington for the weekend in 2019.

I had asked for a meeting at the headquarters in Pequannock because I had wanted to get Breyer's official take on the world that had grown around it. Did they have an inkling from where this universe of competitors had sprung? Why did these replicas of horses have so much allure?

The answer, even for acetate, was familiar: *Because horses.* "They're horses. They are full of character, full of spirit. Somehow they have enough character to draw somebody in," Macejko told me, adding with some defiance: "What model train has that?"

Plastic pintos, palominos, ponies, Percherons, Paso Finos, and every other equine in between piled the tables inside the Berks County 4-H Community Center, frozen mid-gallop, -prance, -graze, -rear, -buck, and -trot on a morning in rural Pennsylvania. Correction: the Breyer horses were not on tables, I was soon to learn; they were in *stables*. Each table bore names like "Harmony Hill" and "Pleasant Acres," demarcating each as a model horse "barn."

As I stepped into the cavernous building, a tinny announcer's voice crackled over the loudspeaker: "English hunter/jumper performance horses to Ring 1, please. Your class is about to begin." All around me in the 4-H center were dozens of adults and children playing with plastic ponies that winter morning in Leesport, Pennsylvania; I had stepped into the world of my childhood imagination. I was at a Breyer competition. And it was real.

Journalism's great gift is demanding that you strive to inhabit the perspective of another human being. But that morning in Leesport, I walked into that room in violation of that precious tenet. I wasn't curious, as a journalist must be; I arrived perplexed: adults toting buckets of bubble-wrapped plastic dolls to obscure corners of the country to vie for which plaything is best? Even the framing, modeled after a horse show with classes and heats, made me silently scoff, "Performance hunters?" In what way was a static figurine performing anything? Some horses were posed in a diorama of jumps, but most baffling to me were classes where the horses were shown "in hand." In the real horse world, that's a type of competition where horses are displayed like show dogs are, paraded in the nude past a judge who analyzes their physique. In Breyer world, it's lining up identical figures side by side, and somehow a winner is chosen among the clones.

To my right, a woman jumped up from her folding chair, evidently flustered. She raked a hand through her close-cropped hair, snatched a black plastic horse from amid the herd of about one hundred in front of her, and scuttled off. At "Ring 1," a folding table in the middle of the room, she hastily plunked the figurine in the center. "Don't start the class yet!" the woman, Mary Ann Snyder, beseeched the announcer, "I have to groom him!" Out of the back pocket of her Wranglers she fished a pink makeup brush and quickly swiped down the hard plastic hide in meticulous swirls in the same way I might apply blusher.

Snyder was sixty-eight years old. Of the fifteen hundred Breyer horses she owns, about one hundred or so were set before her in rows ("congas" in Breyer parlance) of identical models in different colors inside the 4-H center. They were just some of the show team she has been competing with at live model horse competitions for over two decades. More were still packed away in fuzzy pouches, traveling cases labeled individually with the name she had given each plastic animal.

The judge had just called the results over the loudspeaker, and Mary Ann grimaced: her horse had not placed in the first class. But she brushed it off quickly; there were many more heats to go and scores of other models she had brought to ready for their divisions. She gathered the Breyers that were done for the day carefully. One by one, she plucked each carefully from the ring and returned them by the armful to their bespoke sacks stowed under her own folding table, or, rather, *stable*.

She wore a white sweatshirt with five horses head-on, as if they were galloping off her chest, and as we spoke back at her "barn," she held identical Andalusians in each hand, pointing out each subtle difference and sculptural nuance the way a handler of actual horses might revel in bloodlines and pedigree.

After four hours in the 4-H center, surrounded by cavalcades of plastic horses marched by their owners to and from competition "rings," I still couldn't see the difference. Maybe I wasn't letting myself see.

Judges were drafted from the ranks of current and former competitors. But what were they judging? Sometimes it was the composition and scenery. A few were displayed in intricate dioramas like the ponies I'd pose in the Kalahari of my sandy backyard, but others were just lined up on tables, one store-bought model against another. At the Reeves warehouse in New Jersey, I had learned that while the Breyer horses are mass-produced, the finishing touches for each one are applied by hand. That means that while to the uninitiated every dapple gray looks like another and another and another off the same assembly line, there is a world of difference to the serious model collector.

For true enthusiasts, no paint job is exactly alike. To their expert eyes, some have almost velvety touches, brushstrokes that ripple like muscle; others have an eye that sparkles or a whorl of hair daubed just so. Judges look for horses that have the sharpest paint job—things like no bleeding edges in a pinto's spots or the softest marbling in a bay's dapples. The ineffable quality of being almost real. But for the life of me, as I stared at what seemed to be identical toys, I couldn't tell the difference.

Mary Ann keeps a rotating exhibition of four hundred of her horses on display in her living room in Carlisle, Pennsylvania, she said, with a few of her husband's coins from his collection wedged in between to keep him happy. More are in her basement, but the ones on display are draft horses, her favorite type to collect. But outside of her home and the walls of buildings full of the like-minded like the 4-H center that morning, she doesn't speak of her passion for putting model horses on tables to be judged. "Some adults collect dolls, or action figures, but of course they all have to say: 'Oh, they're all in the original boxes, and I keep them because they're going to appreciate in value.' People understand that," Mary Ann said. "Do any admit they play with them?

"*This* isn't acceptable," Mary Ann said, brandishing the Andalusians in her hands to gesture with their hooves to sweep in *this,* the entire room. It was packed with girls and women mostly, whisking flecks off horses with makeup brushes or doing a brisk business of trading rare models between each other. In one corner, a tween in a sparkly scrunchie and a line chef (who confided in me earlier that she had called in sick from work to attend the show) haggled over a filly with ferocity I'd seen only in a Tel Aviv souk.

"This isn't acceptable," Mary Ann continued, "outside this room."

It was a world entirely at home with itself. Obscure or odd to outsiders, no one there could care less. All afternoon I strolled the room, struggling to see the models as the people who loved them did. I squinted at the faint differences in paint jobs on a gray's dapples that made one a winner and a seemingly identical one next to it a runner-up. I puzzled over what exactly it was about the daub of pigment in the eye of the Arabian champion that was superior to the blot in the Akhal-Teke that the judge had overlooked completely.

Competitors that afternoon shared with me the roots of their obsession. For most, it was the fact that real horses were financially inaccessible. Breyers began as a substitute and morphed into the passion itself. "I deliberately avoid real horse shows," Mary Ann told me. "They make me too wistful." But how could the unloving plastic, the toys that perform nothing at all, I wondered, do anything other than make her pine more? Over the folding tables piled with figurines stretched the shadow of not-horses.

Even as competitors cooed over Clydesdales or groomed the dust from the grooves etched into a Bashkir Curly, the totems treasured by every one there remained to me as lifeless as ever. My imagination, the ability to see a gallop in the resin, shut off like a dry tap as I found myself doing what Mary Ann Snyder never did: I stood there thinking about what everyone else would think.

At one end of the hall was the lone male competitor of the day—a sixteen-year-old boy with one green eye and one brown. Devon Frinzi sat cross-legged on the floor, prying four identical models of a gray horse from their yellow boxes. It was a new model released that day, and Devon's father had ducked out from the competition to a local tack shop to buy him four editions, now more costly than when I was a kid, at thirty-eight dollars a pop. As if he wore a goldsmith's loupe, the boy peered over every dapple and seam of each clone, seeking to determine which paint job was superior. He would keep the one he felt was best; the rest he would resell, a business that, according to his father, Dave, a schoolteacher, earns his son about six thousand dollars a year. Devon promptly spends it on more Breyers, of course.

"His twin brother is this number-one-stud-type kid, trivarsity, all the sports," his father told me as he unboxed a figurine for his son. "That's just not Devon. *This* is Devon." Dave opened his phone to show pictures of the father-son road trip to BreyerFest, a ten-hour drive from their home in Easton, Pennsylvania, and snapshots of the pair camped for two days outside the Kentucky Horse Park to be the first through the doors when the convention opened. Devon was a near celebrity among the other Breyer devotees in Kentucky for his encyclopedic knowledge of the Breyer catalog (and because there are hardly any other boys). Other boys play *Grand Theft Auto*, not with plastic ponies. "I don't care what they think," Devon told me. "That's not me." This is Devon.

"I thought I would just sit here at these competitions," his father said. "But his thing became our thing. I wouldn't change it."

Then I realized: all the people milling around the 4-H center that morning had the ability to see in the figurines what I no longer could as an adult—not just the models' intricacy, their subtlety, but their *life*. What was happening in Leesport was what Mary Ann had pointed out is so often forbidden to grown-ups, something we lose as we age—pure, unmitigated *play*.

I turned to watch Devon as he left the pile of yellow boxes to go judge a table full of teensy ponies, 1/32 scale to the real thing, and that ugly emotion I had battled all day—I realized it was judgment, even condescension—fled from my chest. I envied Devon, I realized, a person who did what he loved with gusto. So much of my life with horses had been about fitting in, about looking the part, an attempt on horseback to *not* be the Other. I looked around, I was in a room full of those rare humans who embrace the unique, the unusual, who have the strength of character that I so often lack to not give a damn.

I wished that at sixteen, I could have been Devon; I wished that at sixty-eight, I will be Mary Ann.

From a ten-dollar bin of chipped-up cast-offs, I bought a battered Breyer from a twelve-year-old. I stared at it on my dashboard the entire drive home.

MISTY

The sky opened up seconds before the ponies appeared.

No, that phrase doesn't nearly capture the storm: it was July 2018, and the sky wept, caterwauled, drummed, deluged. It slapped down hot rain on my head so that it cascaded down my back and ran in rivulets down my cleavage to soak me to my underwear. I sat exposed to the elements in my dinky red kayak as it filled with water, blinking away fat raindrops and shaking my head at the absurdity of the situation: it was before dawn, just past 5:00 a.m. I was plunked in a brackish channel off the coast of Virginia, sopping wet and getting wetter. Still, I dipped my paddle into the dancing water and kept on, wending through cattails.

I was after a band of legendary ponies. In a complete and utter squall.

A squall is what brought the band of feral ponies I was paddling toward to their home on the unruly island long ago. Assateague is a thirty-seven-mile-long patch of sand and seagrass swilled all around by Atlantic seawater. The ponies were shipwrecked there, legend goes, cargo on a Spanish galleon bound for Peru in 1550. (No one knows when really. "Once upon a time" is an acceptable date. It's a myth after all.) The herd never made it to Lima; instead, the storm marooned them in an archipelago of barrier islands that make up the Outerbanks of Virginia. Whatever the real origin, what's true for sure is that today 150 ponies, perhaps their descendants, have the run of the land.

"A wild ringing neigh shrilled up from the hold of a Spanish galleon; it was not the cry of an animal in hunger, it was a terrifying bugle—an alarm call. The captain of the *Santo Christo* strode the poop deck, 'Cursed be that stallion!'" So begins a more than eighty-year-old book that I would clutch to my chest in secret under my covers each night as a child, hiding it when Beverly shut off the light in my bedroom. *Misty of Chincoteague*, written in 1947 by Marguerite Henry, stars a butter-and-silver-colored baby pony who once pranced along my bedroom walls whenever I closed my eyes. Henry chronicled the story of the wild foal named Misty, who was corralled and gentled by two local children on the island of Chincoteague, the inhabited island just across the strait from the ponies' refuge of Assateague off Virginia's Eastern Shore. Best of all is that the book is not strictly a work of fiction. Misty was a real pony, and her owners were actual children from Chincoteague. That made the story so much enthralling. *That could have been me*, I thought as I memorized each page. *I could have tamed Misty.*

I was not alone in my infatuation. The tale of the sea-tossed pony was a runaway bestseller in its day, printing millions of copies. It catapulted the real animal into TV appearances and the pages of *Life* magazine and the sandy town of Chincoteague onto the map.

Every night as a child, I held Misty close in my bed as I listened to the hum of the fridge, tucked in what was a former maid's room in our sprawling apartment. It was retrofitted with a Murphy bed that hinged out of a wall of bookshelves that to me were endless shadowed crannies for specters to hide. The years I lived in that room were the most sleepless of my life. That period earned me the nickname "The Wandering Jew" in my family, a tongue-in-cheek biblical reference to the fate of the Israelites to endlessly wander the Sinai looking for the Promised Land.

In truth, every night I set out from the pantry, I was in search of my mother, even though I knew that most often, I would not find her. She was out, still holding forth at a party or a lecture. I would crawl into my parents' empty bed and wait. My company then was the spotted pony Misty, painted by nature with a patch the shape of the United States on one shoulder, her wise eye, encircled by a spot of brown, peering out from the cover of the book.

Every night I fell asleep that way, I dreamed of Misty. Of how she lived wild and free and how a little girl not unlike me befriended and gentled her. I dreamed we played in the sand. The pony was there when my family was not to be found. Horses always were. They always are.

In July 2018, I decided to wander again: this time, in search of Misty.

For hundreds of years, the skittering fiddler crabs have shared the shoreline of Assateague with the band of feral ponies. The equines are as seaworthy as any mariner, sure-footed in tide pools, and canny around marsh muck that can suck them down: "The ponies learned how to fall to their knees, then sidle and wriggle along like crabs until they were well out of it," Henry wrote. Some of the story she fudged. Prevailing wisdom about their origin points to plow ponies loosed from their traces by early settlers to forage here. But Henry preferred the story of the Spanish galleon wrecked on the oyster shoals, freighted with diminutive horses traveling to the mines of Peru who instead swam to freedom. So do I.

But it was not a mythical shipwreck or ponies who can crab walk their way out of quicksand that makes them special, or what drew Henry from her home in Milwaukee, Wisconsin, to the island to tell their tale.

It is because the ponies swim.

In my kayak, I shifted my poncho slightly and was rewarded with a drenching wash from the water that had been pooling atop the oilcloth, now pooling in my lap. The sun was coming up, and I looked for rainbows through the sun shower—anything to distract from the torrent—but there were none. I was float-ing in a narrow stretch of water between the inhabited town of Chincoteague and the wild island of Assateague. I pulled up my hood and stared at the nose of my red boat, bobbing where I had moored it in a stand of submerged salt grass at the center of the Assateague Channel.

There was a commotion on the far beach. The Saltwater Cowboys had arrived. A red barge hung with buoys pulled out from a dock jutting off Chincoteague, laden with a curious cargo: bobbing merrily along in the downpour, a half-dozen horses stood patiently on the boat underneath their mounted riders. In cowboy hats, men and women sat astride the animals as the ferry floated the herd out into the middle of the channel. The infantry had entered: these riders and these horses would round up the wild ponies.

The 150 feral ponies of Assateague Island are the property of the Chincoteague Volunteer Fire Company, a force whose shield is a pony rampant on a sea-blue background. For more than ninety years, the animals have been rounded up once a summer and herded to the mainland for a veterinary inspection and auction. There, the newest foals are sold off to the highest bidder. The tradition is how the municipality maintains pony population control. Every year, most of the spring babies are sold and the rest returned to the island.

Parting baby horses from their dams can be heartrending to witness, but it's part of how a horse grows up. A 2010 study of plains zebras showed that mares wean their foals between about eight months and a year naturally. In the wild, weaning is a less abrupt process than the salty swim and sale, but it can be tinged with violence. There are feinted kicks and nips from mom as a filly with grass-ready teeth goes for her long-suffering mother's teat.

The proceeds from each filly and colt go to the fire company, keeping the small force stocked in helmets and hoses. And for one week a year, the firefighters and their descendants form the corps of the Saltwater Cowboys, tasked with herding the ponies annually across the watery divide. "Hoss Penning Day," as it's called, was the spectacle that first drew Henry here in 1946. It has swelled the seaside town with waves of visitors each July after Henry's story about Misty blew the legend wide.

I watched from my flimsy craft as visitors thronged the shoreline and party boats crowded the narrow canal. A long string of bunting had been stretched from Assateague to Chincoteague so that it spanned the whole channel, demarcating a lane that the ponies would soon splash across. Policemen on Jet Skis churned the surf between the ships, corralling misfit spectators like me who inched their vessels ever closer to the ponies' path.

I steadied my kayak on the wake of the barge of horses, its sides painted with "The World Famous Saltwater Cowboys," as before me it docked at a hummock of sand in the middle of the channel. The Saltwater Cowboys' mounts stood stoic as their vessel bobbed and weaved. When it came to a rest, cowboys and their horses splashed down off the deck onto a submerged sandbar. It would be a way station in the middle of the channel from which the Saltwater Cowboys' horses could dart after the ponies that would soon pass by. With a whoop, a puff of peach smoke zipped into the wet summer air above them, bursting from a flare gun, and the crowd began to roar. The pony swim had begun.

The sight of 150 horses coursing through the water is not something you describe; it's something you feel. I could tell of the way the surf rose to lap their furred bellies, kicked up in broad sprays by six hundred hooves that churned the water white and pawed at the sand. I could tell of how halfway across the channel, they pitched, as one, headlong into a trench hidden beneath the waves, so that suddenly the hock-deep water was up to their throatlatches. And how they pressed their noses upward into the rain-thick air and whipped their legs out of sight under the water so that they were suddenly swimming, galloping still beneath the surf. And how tiny foals with trumpet noses and peach fuzz manes paddled alongside their dams, pale palominos gone dark with the wet, bewildered eyes gone saucer-wide, each fixed on its mother's rump as it dwindled to an island of fur under the tide. But a description doesn't get at how as the ponies cascaded through the water, they thundered into a dream.

All around them were the Saltwater Cowboys riding horses that were sons or daughters of the wild ponies. They were the progeny who had been snapped up at auction and tamed in previous roundups, and they charged at their wild relatives through the water in bursts. The riders flung coils of rawhide whips into the air, snapping them down with cracks that whizzed across the water. At each electric crackle, the ponies shot forward, though the whips never touched a hair. Spectators in bikinis, clutching red plastic cups sloshing mimosas, hung from the flotilla of boats along the route, shouting the ponies onward.

The animals came in waves, herds of families presided over by stallions like Riptide, a silver dapple with a rust body and Rapunzel hair. Like any other celebrity, Riptide has a Facebook fan page, one of many created by horse enthusiasts who document every island pony's pedigree. Ever since Henry, the ponies are so studied by amateur equine genealogists that their lineage can be traced and looked up in breeding booklets, pictorial family trees that sell out each year at the gift shops on Chincoteague Main Street.

The waves of their crossing lapped my cherry-red bow. As the animals were partway across the channel, suddenly four broke away. The quartet wheeled and swam back toward Assateague. They were fugitives like Henry's Phantom, Misty's mother, a mare who capsized boats in her fight for freedom. The renegades swam for the far shore as a cry of "Escapees!" went up from the crowd. I watched as four tails fanned out in the water. Twirled in the long strands and surf was a fifth sea creature: a dark black foal urgently paddling in the breakwater of their flanks. I lost my paddle in the marsh briefly then; before I realized what I was doing, I had cupped my hands around my lips and yelled to the little one: "Swim free!"

In the quiet after the ponies passed through and trotted off on their way to auction, I rowed through the morning, past cormorants hanging their wings out to dry and a snowy egret with a little hat of feathers. I rowed furiously to shore, bumping the nose of my little boat into the dock. Up on land, I wrung out my sopping shorts in the parking lot of the tiny quay, darted into my rental car, and whipped into town to get to the pony parade.

In the wan morning light, hundreds of men and women already lined Main Street. Some kicked back on folding chairs they had set up there hours before, sipping Bloody Marys. Little girls clutched stuffed ponies, peering down the avenue, listening for the clop of hooves. Dotting the crowd were Mennonites, whole families in matching overalls, each large family dressed in the same denim and brightly hued homespun shirt or dress, from toddler to father. Mennonites eschew modern technology as a religious practice; to get around, they rely on horsepower. Most people in the crowd were spectators, but the Mennonites were shoppers, eager to buy the wild ponies to make into riding mounts. Their children would use the ponies to rove the farm the way other agrarian families might use an ATV.

When the ponies finally marched down the avenue, gone was the majesty that had crested from their necks amid the waves. They were small, as ponies of course are, but dwarfed and dwindled by their no-longer wildness. They plodded en masse through town, past motels and delis with the Saltwater Cowboys, whose numbers had swelled to dozens, on either side of the herd. The cowboys pushed the beachball-round mothers trailing babies along the road, keeping the horses away from the tasty, freshly mowed grass of the suburban yards on either side of the street. Their manes were bedraggled, and there was no wonder or even startlement in their eyes. They were simply wet, tired-out ponies in need of a graze.

Past a bronze statue of Misty erected in 1997, the ponies made a sharp right, and marched alongside the Tilt-A-Whirl and Ferris wheel of the carnival past Peaceful Lane, set up in celebration of the penning. They trotted alongside a carousel—painted ponies passing painted ponies—and into a pen at the back of the fairgrounds, muddy from the rain. The gate latched between the ponies and their freedom, and they immediately shoved muzzles deep in waiting bales of hay.

"They get cut up on oyster shoals," said Dr. Charlie Cameron, as he strode into the pen in gumboots and powder-blue coveralls, looking for injured animals. The veterinarian has administered to the animals for the past twenty-nine years, including several times yearly for worming and vaccinations that take place on Assateague. His constant ministrations explained their docility. None even threatened a kick. For "wild" horses, quite a few let me pet their noses as I traipsed in the muck behind Dr. Cameron, surreptitiously reaching out to fondle fluffy muzzles.

There: a bright blotch of red marred the light gold of a palomino's fetlock, or ankle. He was bleeding.

"Hep, hep, hep," Dr. Cameron called, waving his arms at the yearling and herding it with his body down a chute of wooden fence posts at the edge of the pen and into a shanty of medical stalls. A vet tech and a trio of young men, strapping and paunchy, shoved the gate closed behind the shuddering animal, trapping it in a six-foot-square stall. Dr. Cameron rifled around the back of his truck bed and pulled out a syringe. "For tetanus," he said, wiping his silvered hairline with the back of a gloved hand, replacing the sweat on his brow with a smear of mud.

The largest of the young men swung over the fence and into the tiny pen, where the horse quivered in a corner and seemed to grow as small as an English setter. He lunged at the yearling and wrapped the entire animal in a bear hug. The veins on his forehead bulged, as with his sheer brawn, he held the animal to the spot as it bucked and charged to get out of his grasp. Dr. Cameron plunged a sedative into his hide.

A few minutes later, the palomino had been administered the vaccine. He now wore a purple elastic bandage around one ankle as he stood blearily in the corner, recovering from the narcotic. Even under his drooping eyelid, he looked at me with Misty's same wise eye.

Animal activists say the tradition of Hoss Penning is an act of cruelty and unnecessary. In fact, there's really no need to swim: there's a bridge, built in 1962, that the ponies could easily trot over. "The pony swim is cruel and dangerous to the animals and should end now," Kathy Guillermo, the senior vice president for People for the Ethical Treatment of Animals (PETA), told me when I asked for her take on it. I thought of the bright blood on the yearling, of the drooping necks and deflated gaits of the parading ponies. Of the several animals who have died in past swims from exertion, from illness, from stress. A handful over decades, but, still, too many.

"The pony swim is similar to other festivals, such as turkey drops and greased pig wrestling, that have taken their places in the annals of history as examples of the callous disregard of other living, feeling beings," Guillermo said to me. But it had moved me so deeply to read about them and then to watch them surge into the sea. Was it wrong? Was I?

"While these ponies mean a great deal to the fire company, the town, and the county financially, we are also human beings who see these gorgeous animals as the beautiful creatures they are," the fire department wrote in a statement in response to criticism like PETA's. "We handle them with the care and respect they deserve."

After the swim, I found myself sitting on the whitewashed porch of the Conklin family home just outside the fairgrounds. I had asked to borrow their garden hose to slosh the mud off my sneakers, and got an earful about Hoss Penning history. Grandma and Grandpa Conklin sat in rocking chairs, discussing with their fifteen-year-old granddaughter, Hope Abell, which pony she was planning to buy at the auction the next day.

"Go inside and show the visitor what we made you," Carolyn Conklin told her granddaughter. Hope shot her a quizzical, teenaged look and a skeptical millennial "Okaaay?" She disappeared into the house and came out a few seconds later, sputtering. In her hands was a tiny Christmas tree, festooned all over with dime-sized pink and white plastic ponies. "Happy Pony Penning Day!" read a papier-mâché pennant fluttering from the top of the plastic spruce. "Grammy, Poppa, you crazy," said Hope.

"Wherever you are in the world, if you're from Chincoteague, you come back for Pony Penning day," her poppa, Richard Conklin, told me, creaking the rocking chair back as far as it would go. "It's our holiday. It's like religion," he said with a guffaw. There's a Christmas Eve–style dinner, dumplings, turnip greens, and pot pie, the traditional dish. "But instead of shredding the wrapping paper on a pile of presents the next morning," Richard said, "there's a baby pony to buy at the auction."

I considered converting.

Hope raked quahogs and razor clams from the Assateague Channel as an eight-year-old, using the money she earned selling her catch to local fishmongers to purchase her first pony, which mostly go from about three thousand to seven thousand dollars. Hope trains them and sells them as show jumpers. She struggled to explain to me how she selects which pony she will take from the seaside to the show ring, stuttering something about their parentage (thanks to the online obsessives, pony ancestry is on the Web), the broadness of their backs, and the straightness of their legs, fumbling for the words.

"You want them to have just . . . flash," she said. Then she gave in: "When I see a pony, it's like I feel it," Hope said. "I know which one is meant for me. I know, that sounds weird." (Clearly Hope did not know to whom she was speaking.) This year, she said, her heart was set on the little black rebel I had watched swim back. His prison break had been short-lived; Saltwater Cowboys had caught up with him that evening after all, she said, and ferried him—over the bridge, by horse trailer—to the pony pen.

I thought of how his dash for the marshland of his birth had called to something deep in me, something that also wanted to leave all the rules behind and how I had urged him on across the water. What is it about vistas that makes humans want to own the land they sprawl across? What is it about seas that makes humans want to cross them? What is it about wild things that makes humans want to make them their own?

And what is it about me—that all my life I've sought to saddle something that so symbolizes what it means to be free?

"Five hundred dollars, five hundred dollars, do I have five hundred dollars?" the auctioneer's patois rat-a-tatted across the fairgrounds the next morning. He was perched in a crow's nest above the grounds, calling out prices of pint-sized ponies to an arena packed with bidders. In a staging area before the crowd, handlers tugged and pushed foals that had never before been touched by humans into the center of a small arena. The handlers were assigned two or three to a pony, depending on its brawn. They put their hands around its rumps and encircled its shoulders, bracing each to stiffly walk. Every so often, one or another of the animals would let out a shrill, flute-like whinny, and the crowd always responded with a ravished, "Aww!"

A tall firefighter strode into the arena beside a blonde bit of fluff, no bigger than a Labrador. "Just a month old, folks, this one is a fall pickup, returned to the wild to hang out with mom until the leaves are off the trees and she's ready to go home with you," the auctioneer called out. "Can I get a thousand dollars on this little gal?" Before a backdrop of bumper cars and carnival games, the wobbly pony wouldn't budge. Wrapping his arms around her, the fireman heaved the entire pony onto his chest and proceeded to promenade around the arena, carrying the horse the entire time as her legs feebly pinwheeled in an aerial rear. He paraded her from one corner of the stage to the next, as the crowd cooed and craned to see the pony cradled in his arms. I sat on my hands. The cuteness was unbearable, and she looked about apartment sized. I was desperate to bid.

Some of the Chincoteague pony foals stay free for eternity. They're called buybacks, and while some foals are culled and sold off to tamp population growth, others are designated to be purchased as a charitable donation to the fire department and returned to the wild forever. They mill in the pony pen with a length of cord loosely looped around their neck, indicating their status. These foals are chosen at random. In the auction, they fetch quadruple the price of the other ponies. The fee is driven up by consortiums of local residents, like a group of women known as the Buyback Babes, who pool their money to keep their ponies home.

As I patrolled the pen with Dr. Cameron earlier that morning, two sisters had stood staring at a sorrel filly with a white star, a baby identical to her in every way sipping from her teat. She was a buyback from six years before, now with a foal of her own.

"That's our horse, Dreamer's Gift," said Anna Beer, a seventeen-year-old from Clarkston, Michigan, standing next to her sister, Amanda, sixteen. "She lives in the wild."

The sisters had purchased Dreamer when they were eleven

and ten years old, respectively, they told me, as we hung over opposite sides of the fence post, chatting. They've come to every penning since to see her, and in her spare time, Anna writes self-published books about Dreamer's adventures. They were for sale at the Eclectic Beachcomber gift shop on Main Street, and the sisters spent the hours they were not staring at Dreamer in the corral signing the books with sparkly pens. When I met them, they were mooning after the pony for which they had paid the fire department forty-five hundred dollars to set free. But to the sisters, Dreamer was theirs, and they spoke of her with all the ferocious love of horse girls who have a pony that lives in the backyard, not one that lives cropping beach grass a twelve-hour drive through five states away.

"We had wanted a horse our whole lives," said Anna. "We had this opportunity to own a horse, and we don't have to pay any vet bills, we don't have to have her feet done, but we could still have a horse," she said breathlessly. A camera with twelve hundred photos of Dreamer in its memory card hung around her neck. "It was this crazy thing for us, to have a horse!"

Who was I to tell her that she didn't have a horse? That what she had was something closer to a star, those ones you get to name for $19.95 through a website, a red dwarf or neutron star barely a flicker on the horizon that you pay to call your own? How could they believe they had a horse when they had never brushed her and couldn't pet her? She would never be saddled and could never be ridden, I thought to myself, my elbows propped on the rough wooden post between us as they prattled on with delight about their pony. How did she belong to anyone other than the wilderness itself?

And yet as I watched the joy radiate from their faces as they mock-complained about how many pictures of the sorrel they would have to sort through back in Clarkston, I was struck by another thought: How was my version of what it means to have a horse—of owning it, of subjugating it—having what a horse *truly is?* The Beer sisters had a horse as it is meant to be, I realized: a wild animal. They are not even permitted to pet her. She is unencumbered, untouched, pure.

The girls had stroked the red hide of their mare exactly once—the day they gave over the money to buy her wildness forever. Out in the yonder, far from the stable, Dreamer is a glorious equine. The sisters have a horse, I realized, in the fullest meaning of the word.

It was as if Anna heard my thoughts: "She gets to have all those things that a tame horse doesn't; she doesn't have to worry about having to work for somebody, having to do what somebody says all the time." Amanda pressed against the fence post and interjected, finishing her sister's thought: "We bought her her freedom," Amanda said.

"Do you ever get sick of Misty?" I asked Evelyn Shotwell, the executive director of Chincoteague's chamber of commerce, a few days before I arrived.

What I knew of resort towns was limited to the Hamptons, where local residents have an uneasy relationship with the vacationers who descend on the island in high season like locusts, turning their homey haven into a partying playground. Bonackers, as the died-in-the-wool East Hamptonites are called, after the Accabonac Creek that runs through the town's humble hamlet of Springs, make their money off summer sojourners.

The summer people eat in their restaurants and need their pools scrubbed and hedgerows trimmed.

Among those who live year-round in East Hampton, Manhattan outsiders like me were known as "cityiots" (pronounced like *idiots*). I spent my youth working alongside Bonackers at the town's bars and beach clubs, dating their sons and bringing six-packs to their bonfires, desperate for acceptance, and to wear yet another hat of assimilation. As I mucked stalls alongside them, my purest joy of summer was that point in every season when at-first-skeptical Bonackers would see me sweating as I shoveled horseshit, forget I was a cityiot, and invite me to a kegger.

Did Chincoteague feel similarly about Marguerite Henry and the interlopers she brought, who glut their town on Hoss Penning Day? "No!" Shotwell exclaimed over the phone. Misty-philes fill the hotels and buy the souvenir baseball caps with the volunteer fire department's prancing pony logo (I bought two), she explained. The visitors keep the tradition—and the firehouse—alive.

"Plus I can't get sick of Misty," she added, as my eyebrows arched a question she couldn't hear on the other end of the line. "She's still here."

And so she is. Her eyes bulging from a globular snout, I faced Misty of Chincoteague on an afternoon that summer where she stands in grotesque glory. Her hide is stuffed and mounted in a deformed facsimile of the pretty pony she once was, under the display lights at the Museum of Chincoteague Island on Maddox Boulevard. She is a morbid hatchet job worthy of Buffalo Bill, the serial killer in the movie *Silence of the Lambs*, who stitches skins together for giggles. Her antique corpse is sandwiched between displays of tinned fish touting the region's canning industry. It stands beside another taxidermied fright: Misty's child, Stormy.

I left the horror and the museum as quickly as possible, zipping my rental car down Ridge Road to Beebe Ranch a few miles away, where Henry first found Misty. Stuffed Stormy and Misty once were the showpieces of a makeshift museum on the ranch, run by Billy Beebe. He is a cousin of the two little kids who tamed Misty in the book. He prefers to be called Billy King, witness protection of sorts from Misty obsessives who hear his name and catch on to the fact that his relatives owned the legendary pony. For three dollars, King would provide guests an audience with the taxidermied creatures before he donated the stuffed pair to the museum. For one hundred dollars, he'll send you home with a chipped feed pail from which he says the famous horses once lapped.

Misty's name still hangs on a rough stall inside the Beebe barn. It was there I learned her true story, the threads of which Henry had embellished into a legend.

Angel's Stormy Drizzle, a pony spotted like a Jersey heifer, leaned into the barn as King spun out the truth. Drizzle is a sixth-generation descendant of Misty, and King interrupted his yarn to shout "pretty leg" at the pony, whereupon she obediently extended a foreleg and shook her hoof. He blew a raspberry at her, and she darted her own tongue out the side of her mouth in response.

"Misty was not a mystical wild pony, and my cousins never caught and corralled her like Henry wrote," King confessed after Drizzle's little show. "She was a domesticated, ranch-born animal." So too was the "untamable" Phantom, Misty's supposedly boat-capsizing mother; that mare was in fact a backyard pet. "My family signed away their life rights to Henry's publisher for just a dollar," he said. "Oh, and a 'thank you' in the book's back pages. That was nice, I guess." As King tells it, that was the sum total of the money they ever saw from a book that has sold millions of copies.

There in the stable, King proceeded to list the tragedies that had befallen the Beebe clan. There were the car crashes that picked off family and illnesses that had gotten their horses. I gripped the edge of Misty's stall as my childhood came apart in the barn. King wasn't finished: Phantom Wings, Misty's first foal, the one whose birth in 1960 was a local holiday for which the town's schoolchildren were let out of school, got into some cow feed. He swelled up and died. He was four years old. I felt sick.

A few weeks before Hoss Penning in 2019, another tragedy befell the Beebe family: the original barn I had visited, where Drizzle blew raspberries and Misty had eaten her oats, burned to the ground. The animals survived.

That afternoon in the barn, the year before the fire, I thanked King for the tour and steadied myself with a ruffle of Drizzle's forelock, praying she would stay forever unstuffed. As I turned to leave, I pulled out my wallet to buy a Misty pail. I paused. The pony I loved was not the one who had eaten oats in that barn. Neither was she that zombie pony in that museum down the road. I put my wallet away. Instead, I asked him to sign a copy of *Misty of Chincoteague*.

I like that story better.

BIRCHBARK

Sometimes I think that Claremont Riding Academy was as much a fiction as Misty's story. If I hadn't ridden there myself, I would have never believed such a thing could have possibly existed. But there it was on West Eighty-Ninth Street: a vertical stable in the middle of New York City.

It was where I rode in the wintertime, after school, tugging on jodhpurs under my uniform skirt and swapping Mary Janes for paddock boots on the M86 crosstown bus. Claremont was built in 1892 and was four stories up and one down, every level packed with horses owned by the riding academy. The horses accessed their stalls via a series of wooden ramps that served as equine staircases to the upper levels.

All day long, the eighty-eight horses who were tenants of 175 West Eighty-Ninth clambered up and down, the scrape of their hooves echoing through the ceiling over the tiny riding ring on the ground level. The stables looked just like any other modest apartment building on the street, but in fact it was purpose-built, equipped with antique air shafts to ventilate the horses, running through the building.

It was operated by Paul and Nancy Novograd. Nancy Cohen had been a corporate lawyer, but when she stopped by for a riding lesson, Claremont altered the course of her life. "I was studying for the bar and I was really depressed," she told me recently. "So I went for a ride at Claremont. Paul was at the front desk, and, well, we fell in love." Today she is an animal talent agent, running All Tame Animals, an agency that supplies horses for movies and the stage. Surely, I pressed her, she must have dreamed of some sort of urban equestrian life—as I ceaselessly did—if Claremont drew her in so fully that she gave up the bar for the barn. "I didn't even know this kind of work existed," Nancy told me. "This was a revelation." But her life's true revelation, she added, was not horses—though it became them—but the man behind the front desk.

I arrived for my first riding lesson there at eight years old, and Paul was behind that front desk in the grimy, glassed-in front office. He fitted me with a still-sweaty loaner helmet and pulled up a stool behind a desk as tall as me. On a wall behind him was a motto: "Culpa equestribus non equis," which roughly translated means, "It's always the rider's fault, never the horse's."

Suddenly Paul let rip a strange exclamation: "Birchbark!"

Little of my life with horses ever escaped the barn walls. It was too precious to share with the outside world, this equestrian Eden where I felt understood. I couldn't risk it being ruptured by the opinions of my peers at the rigid girls' school I attended for thirteen years, opinions that as a young girl, and as an adolescent anxious to belong, meant far too much.

At Brearley, where Ruth Bader Ginsberg is an alumna, and

outside activities other than SAT tutoring or therapy were seen as detractors from an Ivy League destiny, I didn't much speak of horses.

Nevertheless, the horses escaped. They pranced illicitly on the underside of my desk lid, where I scribbled them in pink and yellow highlighter. They reared and bucked in the margins of every math textbook as I struggled with the subject and sketched to feel better. They trotted down the halls on Mondays, when I would walk bowlegged to class with pulled muscles and rubbed shins after an intense weekend of riding. Invariably, a cool girl would lob a jeer in a faux western accent: "Mosey on there, pardner!"

And they lived in my backpack. There, a dog-eared copy of *The Girl Who Loved Wild Horses* by Paul Goble hid. The book is an adaptation of a Native American folktale of a horse herder who disappeared with a wild stallion and his band. She would return once a year to her tribe with a prize colt. One season, she stopped returning. In the final page of the picture book, the stallion stands on a hill ablaze in sunset, entwined with a new mare: the herder had at last become a horse, her hide as dark as her coal-black hair.

The K–12 institution reveled in its intellectualism and upper-crustiness, literally: where others had mystery meat and chapter books, we ate focaccia pizza for lunch as kindergartners and, barely out of middle school, read the *Communist Manifesto*. It didn't matter that we didn't understand a jot of Marx: it was the perfect ammo for our parents to shoot across the ramparts of a gimlet at their competitive cocktail parties.

Brearley, the Little Harvard of the Upper East Side, was named after the fur magnate who founded it. The cruel profession felt to me an apt antecedent of the place. Our mascot was the beaver, after the creatures whose plushness made the Brearleys rich. I would daydream of a long-ago time when they schooled like fish in the East River that rushed past my classroom windows—until, I guess, Samuel Brearley made them all into stoles.

But the true icon of Brearley was Mehitabel, the fictional name the school gave to the ideal Brearley girl we all were supposed to be. She was the real mascot, not just the poor beavers, and when we were admonished to behave, it was this paradigm we were exhorted to be like. Mehitabel was the sort who wore a headband over her towhead, spoke only when spoken to, and then only quoting Ovid. In Latin. She was not, needless to say, an upstart kid of immigrant Jews. It was thirteen years of my life tortured by the facts that I was (and still am) unable to spell, grasp calculus, stay quiet, keep my bangs as neat as other girls did, or be on time. The idea of finding oblivion in horses as the girl who loved wild horses did, of never coming back to a world where you didn't or just couldn't fit in, was tantalizing solace.

At school, curled on a windowsill in a homeroom overlooking the river passing by the foot of Brearley's redbrick tower, I rarely glanced at the onrushing, beaver-free water. I stared instead at that sunset on the last page of *The Girl Who Loved Wild Horses* and galloped away.

"Birchbark!"

Suddenly, Paul's words were echoing across the cream-brick building, pumped through loudspeakers that seemed to be everywhere. In his hand was a CB radio that he barked into again: "BIRCHBARK!" His voice reverberated through the upper

stories, and in a few moments I felt the ceiling vibrate, a shuf-
fling sound like when the neighbor upstairs decides to move
furniture at night. This was not an armoire. It may have been
three armoires.

A spray of dust slipped down through some chink, sprinkling my helmet.

"Go catch him!" Paul barked. I was bewildered as someone
shoved me out of the office and into the ground-level riding
ring on the other side of the dusty panes of glass. I was at the
foot of the equine stairwell, though I didn't know it. What I
saw was a slanted ramp with no stairs, but every few feet or so
was a two-by-four nailed across its width. I looked up. From
the second story appeared the silhouette of a horse, backlit by
afternoon light. He stood at the precipice, and I stood in shock.
The ramp felt like a sheer drop, a slope worthy of an Olympic
ski-jumping trial.

A horse was coming down?

Slap! And from a dark corner, an unseen handler thudded
a palm across the animal's flank; Birchbark launched down toward me. I watched agape as twelve hundred pounds of horseflesh crept down the ramp, his head level with his knees for
balance.

Suddenly the slick surface of one of his horseshoes skidded
over a puck of manure plopped on the ramp. For a few harrowing seconds, the gangly horse went skittering, butt down,
head up like a sitting dog. His rump tucked under him as he
slid, and for a moment he was coming down the way a baby
scooches down stairs. I gasped, but this was not Birchbark's
first staircase rodeo. His hooves caught on one of the two-by-
fours hammered into the ramp, and his brief plummet halted.

Stoic, he scrambled his hind end up from the floor and seemed to compose himself for a moment. Birchbark then calmly continued his descent, this time a touch more gingerly. At the base of the slope, he stepped into the light toward me. I could see Birchbark's coloring—an Appaloosa, chestnut red at base with a milk-spattered hide. He let me catch the reins of his bridle, and I dug deep in my pockets for all the sugar cubes I could possibly provide.

Birchbark and all his fellow upstairs tenants had adapted by necessity to their weird life. They had learned to throw their legs apart like mountain goats to get a better purchase on the ramp and propel their half-tons of horseflesh up or down at what appeared to be an astounding 70 percent grade. Later, when I became a regular at Claremont, I always enjoyed the moment at the end of each session when I would dismount and let Birchbark back up the ramp. He would skip back up the chute with far more enthusiasm than he came down. I soon realized his ginger descent was mostly a case of the "I-don't-wannas" that nearly every employed horse displays before a workout, whether they're being led out of a field or down a staircase. He'd delightedly step skyward after class, his legs expertly set at odd angles for a perfect purchase on the ramp, fairly galloping back into the ceiling to a waiting bucket of bran mash.

Riding in the ring at Claremont was at first pure terror, a trauma far greater than the vertical descent. The arena was bisected at intervals by a forest of columns, struts necessary to hold up the scores of animals who lived above in the more than a century-old building. As I sat on Birchbark that first day, I quickly learned that riding here was an equine version of Frogger, that 1981 arcade game where a pixelated frog tries to cross a busy highway without getting smushed. I white-knuckled my reins. All around me, horses blitzed around the ring. They wove around the columns, one another, and the trainers in the middle of the arena, avoiding each obstacle with differing degrees of success.

"Quarterschool!" screamed a girl as she kicked a rangy thoroughbred past me. It was the Claremont convention to call out which quadrant of the henge of columns you were riding through. Quarterschool, halfschool, centerline—each arcane term delineated loops of different circumferences and trajectories. I tried in vain to comprehend her meaning on this first ride. "Halfschool!" barked a girl on a chestnut with a dirty white blaze. I was not breathing, standing in the center of the thicket of columns, paralyzed. The thud of hoofbeats came from behind us, and I wheeled Birchbark around in a quick pirouette. "I said halfschool!" she barked again just as a big black draft-cross topped with a woman just learning to canter was launching into the gait. As she turned, so did the girl on the thoroughbred—directly into her path. "Oh my God, HALFSCHOOL!" screamed the woman barreling toward her on the draft.

I closed my eyes and saw flickers of Guernsey soaring overhead. Seconds later it ended (as many lessons at Claremont stables did) in a slow-motion, and harmless, crash, the black horse wisely halting as he and the thoroughbred horse glanced chests. The animals aimed a halfhearted nip at each other like taxi drivers flipping each other the bird, while the instructors of the two different lessons duked it out over whose fault the near pileup was.

Shaken, the two riders called it a day, and the ring emptied out for Birchbark and me to trot around. After a few minutes of deep breathing, I realized it wasn't half bad when it was empty. But little could have caused me to find fault with Claremont: I was a child, on a horse, in the middle of New York City. That M86 bus had driven me straight into the pages of my picture book.

Excursions to Central Park were what redeemed Claremont's chaotic interior, when I could let Birchbark have his head on the bridle path that rims the reservoir. The paths are a relic from the era when the park was newly built in 1857, created by Calvert Vaux and Frederick Law Olmsted, and the tracks added in for the gentry to promenade on horseback and trot to their country houses in what were once the wilds past Harlem.

It always stuns me that this groaning, clanking, thoroughly modern metropolis of mine was built by and for horses. Nineteenth-century New York, the raw land on which Victorian-era urban architects laid the rectilinear grid plan, was a horse haven, and continued to be as the city arose. The 1891 annual report of the Health Department of the City of New York tabulated that 62,208 horses lived in 4,297 stables across the city. Other contemporaneous accounts have said that as many as 200,000 horses were working in the five boroughs.

The city still echoes with horses—and not just the two hundred or so carriage horses who still ply their trade in Central Park, sharing their buckets of grain with portly pigeons on West Fifty-Ninth Street. Hidden on street corners in this modern city are marble fountains with troughs for thirsty horses tucked at their base. An obelisk in Harlem's Sugar Hill neighborhood is surrounded by a tureen from which horses once sipped. Under the Ed Koch Queensboro Bridge a beautiful woman still reclines in an art nouveau tableau of mosaic tile; she sprawls above a sculpted bull's head that spits water into an animal manger. When the Brooklyn Bridge was opened in 1883, the toll was a nickel a horse for the crossing. It's uncanny to me that the width of my city's taxi-choked streets corresponds to the breadth of two horses abreast; after all, horses were yellow cabs when my city was new.

Tarnishing my vision of horses prancing in my streets is one unavoidable fact: where there are horses, there is horse poop. An average-sized horse produces between thirty and fifty pounds of manure and over two gallons of urine every day, according to some unlucky soul at Rutgers University in New Jersey who apparently drew the short straw to study equine excrement.

What to do with the nine tons of manure each New York horse deposited on the sidewalk every year (using that nineteenth-century Board of Health report's horse population count, I calculate that's over half a million tons of crap: 559,872 tons, to be exact) was a topic du jour of a series of 1880 *New York Times* articles and something of a public health scandal: manure piled on street corners, attracting flies, stinking to high heaven, and terrorizing the populace. One manure dump on East Forty-Sixth Street was responsible for "emitting a stench which defies any attempt at description," the *Times* wrote in 1884. Feces fear was whipped up by firebrand reporters who wrote frenzied articles claiming that horse excrement was to blame for twenty thousand deaths a year.

And so, despite the fact that early cars were far more dangerous than plodding horses, automobiles took off in New York City in part because cars don't poop.

The few horses left on the park's bridle paths in modern times were Claremont's. Aloft on an animal's back, I found a tranquility I rarely reached in my urban hometown. Central Park on horseback is serene—except for the beginning of each excursion, which was just mayhem. That's because Claremont was not inside the park but on a side street several blocks away. Every trail ride began with a jarring walk across asphalt, the animals schooling like salmon up Central Park West against a river of cabs that didn't seem to think horses were worth tapping the brakes. "It's just too inviting to pedestrians and dirt bikers and people throwing Frisbees and people pushing strollers, and it's a zoo out there," Paul said in a public radio interview the year Claremont closed. "And our horses are, thank you, just too polite for zoos."

On our first jaunt, Birchbark was more interested in trying to eat the budding tulips that encircle the street trees than in the hurtling cars, thank goodness. He didn't blink at the puff of a lapdog that lunged at us with the bravado of a grizzly. But the horse was careless for his own safety in his eagerness to get to the cool shade of the park, and he tap-danced on the sidewalk as we drew near. I hauled him up as a hot dog vendor scudded his pushcart down the avenue, missing us by a hair. To get away, I steered the horse toward the curb, where a heedless Manhattanite flung open her taxicab door in our path. I swear Birchbark would have jumped it in his eagerness for the green space had I not screamed some version of, "I'm trottin' here, lady!" and she slammed the door closed just in time. Sometimes delivery guys on bikes circled us like buzzards for fun, tinkling their bike bells at us with menace. But Birchbark was mostly stoic. A true New Yorker, he'd seen it all.

If you survived a ride inside that ring in the garage at Claremont, the city streets and tourist-glutted park were nothing in comparison. After weeks of lessons, my college-aged instructor at last deemed me road ready. That afternoon, instead of the terrifying ring, we tap-tapped hooves on the concrete of Eighty-Ninth Street and ducked into the park. Inside, I unclenched my grasp on his reins until I held them just by the buckle, and suddenly Birchbark, starved for greenery in his city stable, let loose.

We careened down a trail past the Dakota, where John Lennon once lived, tossing up woodchips on the bridle path in our wake. It took a few minutes to wrestle the joyful horse back into my control, but once his jollies were out, he dropped to an easy trot alongside the stone park wall. On its far side lay the dark street corner where the Beatle was murdered in 1980. On autopilot now, Birchbark hung a right not far from Strawberry Fields, the memorial to Lennon that lies inside the park near Seventy-Second Street, established by his wife, Yoko Ono. It's perpetually thronged by Beatlemaniacs, and that day was no exception. We pranced past several, strumming, as they always are, a botched version of "Imagine."

"That's a horse!" a startled Beatles fan yelped to no one as we flounced by. "Or this is a really bad trip."

Horses lived on West Eighty-Ninth Street for 115 years, until Paul closed Claremont one afternoon in 2007. Inside the barn that day, it felt like a wake according to news reports. Riders gathered and recalled winter rides when waiters from Tavern on the Green, a glamorous restaurant in the heart of the park, served the horses carrots on sterling platters. Paul had taken out a second mortgage on his house, he told reporters, to keep the academy afloat and was dwindling toward bankruptcy. Central Park had flourished with a revitalized city, and in doing so, it had become too hazardous for his horses, he said. Horses were an anachronism the city had no room for anymore. When he shut it, the horses were given new jobs with All Tame Animals, Nancy's talent agency, or given away. Claremont was sold to the Stephen Gaynor School, a private school for children with learning disabilities.

"I think the community, on one hand, they were sad to see a nostalgic place go," Gaynor's head of school told me when I interviewed him about the move in 2010. "But, on the other hand, they were happy not to have to step into horse droppings anymore."

AMIGO

I stood in the stable aisle at midnight in my pajamas—a Bart Simpson T-shirt and a blue baseball cap pulled backward to tame my sleep-tangled hair. I knew nothing about Amigo, a horse fobbed off on my family for free by another, other than a fact that made me vibrate with joy: *He was mine.*

Soon I would find out that my corpulent red quarter horse was afflicted by a debilitating neurosis: a harrowing, soul-destroying, mind-erasing fear of plastic bags. But as twelve-year-old me gazed at my first ever horse (squeal!), in his fat, short, russet glory, all I saw was a cut-marble masterpiece.

It was dark when my mother had woken me and bundled me half asleep in my rumpled shirt in the Ford Taurus to head to the barn. Amigo had arrived. We drove silently through the woods to the stables. I felt that if I spoke, I might physically pop. I was the girl with a horse! I was the girl I never thought I'd be.

The stable was silent except for the metallic whir of cicadas romancing one another in the oaks above our heads. I flicked on the lights, and down the barn aisle, horse heads snapped up, shocked at the flood of brightness. I walked tembling down the swept cement, afraid that if I ran, the moment would burst like a soap bubble and the horse would disappear into the night. I mashed my cap down on my nest of brown hair and turned to face the small animal in the stall. Two nut-brown eyes blinked back at me, a strand of hay hanging from a thick lip. In the stall stood a dumpy, unbothered equine. Mine.

I was in love.

Amigo was the first of five free horses I have been given in my life. The reality of horses is that as big as they are, they often fall through the cracks of a life. There are new jobs too far from the barn and ugly divorces. Roswell, a bay Oldenburg I once rode, had to be padlocked into his stall in New Jersey each night as his family's marriage unraveled. He slept with a bike chain wrapped around the iron bars, I was told, lest the ex-husband, who had stopped paying his share of the bills, sneak in and slink off with him in the night. One afternoon I arrived at the barn to find Roswell's name tag read "Remy," a sort of equine witness protection.

But perhaps the number one reason horses float around, riderless and forgotten, is biology: the little girl grows up and one day realizes there is something more interesting than her pony: boys. That was how I ended up with my first horse, and my first free horse, Amigo, I was told, the once beloved pet of the daughter of some family friends. Girls grow up, become teenagers, and dump their horses for a dude.

I was almost a teenager, and my parents had moved up in the world. My mother had emerged as a regular figure on television, and I'd cringe over a meal of Bev's ackee and saltfish as my mother leveled Mom Voice at squabbling couples on air. Mom and Dad had churned out their second jointly written self-help book, *Not Quite Paradise, Making Marriage Work*, years before, at the time a groundbreaking guide on how to stay hitched. My father's memoir had followed and was well reviewed, and their shared private practice was thrumming.

Still, purchasing a quality show horse was beyond their means, but they looked at the books when Amigo was offered to us gratis and found, to my shattering joy, that they could support a gift horse. Competition horses have two names, a schmancy show name and a nickname, or barn name, they're called at home. Amigo was just Amigo. But it didn't matter that he was a far cry from a glossy Hamptons show pony, a squat red ball of a horse with an aversion to forward impulsion and a host of soon-to-be-revealed debilitating quirks. It was fitting that I first saw Amigo in the middle of the night because Amigo was my dream.

To my father, Amigo was an Oriental carpet.

In the 1970s, my father was awarded reparations from the German government for murdering his father. The first check was about ten thousand dollars, he told me, which took several appeal processes to obtain, after an initial assessment of his happy American life by a German inspector resulted in a letter that still burns a hole in my mother's filing cabinet. Dr. Nir's growing up fatherless, forged in war, had been character building, it said: reparations denied.

When the check finally was awarded, my mother suggested using it to pay bills or do something useful like fix their rickety old Jaguar that had once caught fire on the Triborough Bridge with me stuck in my car seat by a jammed seat belt. He refused. It could only be used to buy something they didn't need, Dad told her. It would be a statement that he had needed no help, had needed nothing but his own power to survive. The only thing the German blood money could be used on was luxury, Dad said—nothing essential, nothing weak.

They settled on the Oriental carpet that blanketed the formal sitting room we used only to entertain guests on Jewish holidays. The added perk of a carpet, my father would say mischievously, was that he could step on the Germans whenever he wanted.

Wealth was something my father admired not for its own sake or for what it could buy; he admired it as an unassailable marker of our invulnerability against the best efforts of murderers who tried to wipe him out.

To my father, our horse, our Hamptons horse, was proof the Nazis had lost.

"Noli equi dentes inspicere donati," an Italian man named Eusebius Hieronymus, later canonized as St. Jerome, wrote in 400 CE in his Letter to the Ephesians. It is the first-known reference to the admonition, "Don't look a gift horse in the mouth." Nevertheless, before we accepted Amigo, we ignored St. Jerome's advice. We looked my gift horse in the mouth. Thoroughly.

Teeth the size of piano keys line every horse's mouth, fat trunks of dentin and enamel that are vital for hewing down and pulverizing forage. Bunnies and beavers have teeth that are ever-growing and filed down with every chew. Horse teeth are finite, and like the rings on a tree, they can be used to adjudge a horse's age as they whittle away. Fresh teeth erupt at different stages, like wisdom teeth on a human teenager, so the completeness of a horse's dentition can also tell you how old the equine is.

"A horse that has lost his milk teeth cannot be said to gladden his owner's mind with hopes," wrote an Athenian named Xenophon in a treatise on horsemanship about 350 years before the Common Era. In other words, you're buying an old nag, and that, he added, "is not so easily disposed of." (I guess Xenophon never met my spicy Polish cousin.)

"Teeth can tell you a lot about the character of a horse," Dr. Brendan Furlong told me one afternoon. Dr. Furlong is the official vet of the US Olympic Equestrian Team for the discipline of eventing, where horses jump massive naturalistic obstacles at speed. He has traveled with his patients to five separate Olympics, from Sydney to Athens to London. He also happens to be my horses' vet—not to mention cantankerous and punctilious in a way that only people with musical Irish accents and shamanic medical skills like he has are allowed to be. He has no truck with people who delightedly point out his surname is how you measure a racehorse lap. When he examines my horses in the noisome August heat at their barn in New Jersey, a coterie of assistants surround him. They stand in a circle as he works on a hock or a hoof and fan away the flies from around him with rolled-up medical charts like flunkies with palm fronds.

"One of the best things about treating horses is they can't prattle on at you and say, 'This hurts' or 'That hurts,' or be a hypochondriac, like you humans," Dr. Furlong said, his prickly shtick in fine fettle during our interview. "They don't distract you while you're doing your work, and they don't try to influence you by telling you what they think they want you to hear. 'Tis a very honest and straightforward dialogue between your diagnostic abilities and your ability to read the horse."

That silence is also why teeth are so vital as rune stones to reveal a horse's past: what the animal can't tell you, their mouths can. Uneven tooth staining could mean a poor diet. Scars and abrasions or, in the days before microchipping, perhaps a tooth filed to fake an animal's age can all lurk inside the bite. Uneven wear on teeth could indicate a horse has stable vices, such as cribbing. Some believe cribbing is how horses get high. A jailhouse behavior most often born of boredom in the stall, cribbers nip the edge of a hard surface, say, a bucket, and suck in air with a burp. It is thought to release endorphins and give them a little buzz. It's a vice because the chomping can chip away at wood finishes and induce gastrointestinal distress. (Cribbers just think it's fun.)

"The mouth and the teeth can tell you things about the horse's previous handlers, and previous riders, and whether there were unscrupulous things done," Dr. Furlong said. "The teeth can be a window into his past. You *always* need to look a gift horse in the mouth."

But in truth, Amigo could have had a jack-o-lantern smile and he would have come home with me. He was my gift horse, and there was no convincing me he possessed a single flaw. But flaws he had. He was afraid of plastic bags, items that litter every barn because they are a necessary part of transporting horses' favorite treat: carrots. Amigo jumped two feet up and six feet sideways every time one crinkled, bloodying my nose frequently as I catapulted into the dirt. So be it. He was still perfect.

Plus, I understood him. Horses are prey animals; they possess very little by way of offensive weaponry. As big as their teeth are and as stone-strong as their hooves may be, a bite or a front strike is little match to the arsenal of claw and canine possessed by the creatures that historically ate them. They're little match in head-to-head battle with the jackals and lions that picked their ancestors off on the veld, the hungry wolves on the ranges, the pumas on the Andean steppes, or the sabertooths that stalk them through time on the walls of ancient painted caverns like Chavet.

It's called a spook. It's the behavior of shying away from strange and (not-so-strange) objects, and it is an evolutionary advantage for tasty herbivores. Creatures who freak out at something and start away instantly, even before they have deciphered what the scary thing is, survive when the divot in the spear grass is in fact a lioness. Those equids who stopped to carefully consider whether their anxiety at the shadowy patch on the ground was warranted ended up snake food when the rustle was a cobra, never to pass along their sober, ruminative genes to posterity. And so spookiness has persisted in horses. It is a hardwired property difficult to breed out, but it's also useful. It is a touch of terror that makes the Olympic horse desperate not to graze the poles over which it jumps. Productively channeled hysteria is what makes a racehorse blitz its way down the track.

Horses run away from danger for another reason: it is from behind that they are most dangerous. Horses are safest in flight, not just because they are making an escape but because their rear guard carries its best battle artillery: the kick. And while horses can slash with their forelegs, stomp, and brutally bite, they are truly weaponized from behind, flinging hard hooves at a velocity that can cave in a predator's skull.

How hard is a horse's kick? Standard lore is that a punch

from a hoof is the equivalent of being hit by a car going twenty miles per hour, but it's unclear who came up with that oft-repeated metric. (A 2016 study with specially calibrated instruments clipped to horse toes mostly resulted in Swiss scientists' proving the need for better means to take measurements.) I've seen hooves splinter wood like Jackie Chan and pucker sheet metal like a Jurassic Park raptor. I've heard of a hoof to the chest that stopped a handler's heart.

A good horse for the most part won't kick you anymore than it would its mother. Nevertheless, the adage to give their rear end a wide berth is gospel to riders, and we do it unconsciously as a habit. And despite his phobias, Amigo was a good horse, if weird. I gave him a show name— "A Tad Eccentric"— because he very much was.

Looking back, I still can't believe the Amigo years were real. I woke up every summer morning at dawn in our house by the sea. I left my grandmother in bed to take a beat-up blue commuter bus from our still-small-town segment of East Hampton known as The Springs to work at the barn. But the bus was whimsically scheduled, and more often than not, I'd hitchhike.

Busy with entertaining or trolling yard sales for the strange glassware they collected as a hobby by the thousands, my Manhattanite parents didn't understand how suburban life worked and that schlepping your child places is part of the deal—or they just weren't willing to be inconvenienced. Each morning they'd set me free and expect me to get to where I needed to just as I did during the school year via the M86 crosstown bus and screeching subway. And I let them believe that. Nor did I ever explain that in the woods of the Springs hamlet, this was almost impossible without sticking out a thumb and clambering into the first car that stopped for me. I hitchhiked everywhere. Hitchhiking made me feel somehow like Huckleberry Finn on the raft of some stranger's Toyota; a Boxcar child of privilege.

Work was East End Stables on Oakview Highway, a road near the red-brick volunteer fire station in town. There, I tacked up lesson horses and taught pony campers a few years younger than me in exchange for riding instruction for me and Amigo. It was an arrangement that kept my riding partly funded from the time I was twelve to when I was past twenty. In between lessons, I exercised animals like Cotton Candy, or C.C., a prissy dapple gray pony with a pale pink snout. True to Sarah's Axiom of Ponyness, she was not as sweet as her name implied.

One aspect of my job that first summer was to be the crash test dummy on C.C. each morning before the girl who had leased the pony came for her afternoon ride. Full of naughty morning pony jollies, C.C. would buck me off every single time I rode. I recall only one time she did not throw me—when the saddle slipped beneath her roly-poly belly with me on it, doing the work of dousing me in the dirt for her.

Nevertheless, I climbed back on every morning as the pony glared back at me over her shoulder, readying her catapult. I resented my job of tiring out the bratty creature so she'd behave each afternoon for her true master, a pint-sized, raven-haired sprite who looked like Wednesday Addams. I hated the girl passionately.

After about a week of my loathing the girl, we became inseparable, bonded by our mutual addiction to equines. After pony camp was over for the day, the girl and I would ride Amigo bareback and double, her behind, me in front, in cut-off denim shorts, down a small dirt road behind the barn that led to a farm stand. We'd startle posh shoppers from the city, tying Amigo to a fence in the parking lot (out of sight of any plastic bag poltergeists), as if we were in a spaghetti Western and the organic grocer was our local saloon. I always got a mint iced tea, pouring it over my palm to feed Amigo the drink in deep slurps. The farm stand was too fancy for PB&J, so my friend and I would split a loaf of bread and a log of goat's cheese for lunch and tote it from the shop back to the barn tucked under an arm, both of us on Amigo's back.

It was all so hopelessly decadent, but we were naive to the image we cut, prancing on horseback to do our luncheon shopping in the fanciest town in America. We picked blackberries from horseback on the way home, and dismounting back at the barn, we ate hunks of bread and cheese in the shade with Amigo tearing grass beside us, in silence. It was a decade before cell phones, and we don't have a single picture to document our ritual, and I'm glad for it. We were so present I can still smell the grass, the brine of horse sweat, the sticky mint on my palm.

Best of all, it was a world away from the high pressures of high school and my higher-pressure parents, and the whizzing, buzzing city that enveloped us and kept us far from horses most of the year.

It was the only world that felt my own.

BILLY

My father's favorite opera, *Aida*, is my favorite opera too, but for my own reasons: smack in the middle of it is a scene out of a horse girl's dreams.

Dad went to the opera obsessively. As a precocious little boy in prewar Poland, he would sneak in to the Teatr Wielki, in his hometown of Lvov, and inhale Puccini and Mozart from the Beaux Arts rafters. It was a precious time before his life was ruptured by war, the short years before his innocence was stolen. The sanctuary of that velvet and gilt opera house was something he sought to re-create in New York City.

When I was around twelve, he deemed me old enough to go with him. To *Aida*. That evening, my pulse quickened with the timpani, the bass drum, the horn, the harp, the violins that swelled to bursting across the stage. The Met's golden curtain rose on ancient Egypt. I gripped my father's hand. Act II: Verdi's "Triumphal March" had begun. Under the starburst chandeliers of the grand opera house strode a victorious Egyptian army, returning from war with boundless booty. They sang my father's motto—"*Ritorna Vincitor!*" Dancers leaped carrying barrels of gold bullion, dancing duets with the glittering armor of their enemies. Warriors herded captive handmaidens dripping with silk and whipped along their slaves—muscled enemies now bound in chains.

And then, right there in the Metropolitan Opera, smack dab there on Manhattan's Upper West Side, to my shock, glistening horses strode through the gold curtains and across the stage. The first time I saw them, there in the rapt opera hall, among the blue-haired ladies and bow-tied stuffed shirts, I shrieked.

It was all I could do to keep from shrieking more than twenty years later as I stood beside that same golden curtain—but this time on its *other* side. It was just before showtime on an evening in 2019 and the Met had given me permission to observe the equine stars of Dad's and my favorite scene in action. To steady my excitement, I ruffled my fingers through the striped forelock of a Norwegian Fjord horse named Billy. He was standing backstage beside me, waiting for his cue.

In seconds, that curtain would rise for the splendid Triumphal March—and Billy's trot-on part. But first, the clarion call of Verdi's French horn sounded. Billy's floofy ears perked up. He lifted his shaggy blond head and sniffed the air. All around me and the palomino, in the dark of Lincoln Center's backstage, lithe actors draped in ancient Egyptian tunics scurried on sandaled feet to their places.

And then, there in the wings of the Metropolitan Opera's grand stage, seconds before showtime, the little horse peed.

"He always does when he hears that first note," Lori Allegra, his owner, whispered to me. Fortunately, she stood backstage with us, armed with a bucket, and never misses Billy's signal—a tiny stretch to his tiptoes that precedes the flood. "He's excited for his grand entrance. Wouldn't you be?"

Backstage, amid the kaleidoscope of music, art, and horses, memories of my father came to me sharply, like the clash of a cymbal cutting through an overture. I ducked behind a beam, narrowly avoiding collision with two dancers as they skipped offstage. They stopped after their perfect pas de deux, turned to each other, and surreptitiously high-fived. I tried to make myself scarce in the tumultuous goings-on, both to compose myself as loss and awe rocked through me—and to avoid messing up the clockwork pageant unfolding all around me. A thick lump shot to my throat for how viscerally I wished my father were still alive to savor this secret pocket of our city. This had been his world, and it hurt, as it does with all people who leave us, to know it kept going on just fine without him.

An elderly man in thick eyeliner sidled up beside me and the horse, tugging on a bald cap, transforming before my eyes into an ancient Egyptian priest. He idly scratched his foot with his ankh-topped stave. "Is this your full-time job?" I asked him in a stage whisper. "Oh no," he replied as he skipped off to make his cue. "That would be crazy. I'm a psychiatrist."

Billy and his identical teammate, Bobby, champed on their bits placidly all the while behind a sound-dampening curtain. They were hitched to a chariot, in which they would tow the great returning *vincitor* Radamès before the king. The pair were more interested in carrots than cantatas, and as the great commotion of the opera corps prepping for Act II swirled around them, the two dun-colored horses stood serene as befitted their Nordic breed. Every so often, one nose or another would gently snuffle the passing troops of the Egyptian army just in case there were treats hidden in the folds of their sarongs.

The opera's equine green room is the underground parking garage beneath Lincoln Center. We had started the evening in the subterranean lot, where I was surprised to find a horsewoman from my childhood: Nancy Novograd, who had run Claremont Riding Academy, staggered up to me, weighed down under an armload of spangled vintage horse costumes. Claremont had supplied the horses for *Aida* since 2000. After closing Claremont in 2007, Nancy and the handful of horses she kept were still doing the job, now through her animal talent agency, All Tame Animals. All Tame has provided everything from orangutans for *Vogue* to Russian Borzoi for the American Ballet Theater. The stage-trained horses now live placidly on her upstate farm, their occasional cameos a far cry from the chaos of the Claremont riding ring.

Nancy glanced around impatiently in the parking garage. The production would include four horses that night: two glossy bay geldings, Nacho and Córdoba, who would be ridden, and the two Norwegian Fjords, Billy and Bobby, who pulled the chariot. Upstairs in the lobby, the little xylophones opera ushers plink to signify it is time to take your seats were sounding, but the horses had still not arrived. Suddenly the unmistakable clop of hooves interrupted Nancy and my excited catch-up session, incongruent echoes in the parking garage.

Nacho and Córdoba emerged from between a red Mercedes and a mint green Honda, each led by handlers, stretching their legs in the few minutes before showtime. Nancy hurriedly ran between the geldings, affixing sparkling fringes, beads, and baubles to their bridles. They would precede the chariot onstage, bearing warriors on their backs. Within minutes, the pair were caparisoned warhorses, spangled and tassled, befitting the kingdom of Radamès. Nancy fished in her pocket and pulled out four balls of fluff that she popped in each of the horses' ears. The stoppers would deaden the soon-to-be crescendoing music of the Triumphal March and, the hope was, any horsey nerves. The final touch was a single blue ostrich plume. She pulled two from the back pocket of her jeans and spiked the feathers into a little holder on each horse's brow. They dipped and bobbed with aplomb. Her work done, Nancy smoothed Córdoba's forelock.

"It's showtime!" she whispered to him.

Billy and Bobby were already dressed in leather harnesses. They had their own handlers, their owners, Lori Allegra and her husband, John, who had tethered them beside the stage door as they busied themselves costuming the horses. Lori swung Billy's emergency bucket in one hand, at the ready. John threw a gold lamé blanket fringed with wooden tassels over Billy and knotted on his own sarong. John performs in the opera too—the essential role of nose petter, keeping the pair steady while Radamès sings from his chariot perch. Novograd had contracted the Allegras to provide the Fjords: the couple collect antique carriages and the horses who pull them.

As we gathered in the parking garage, readying the small herd for the stage, I realized with a start that I had met the Allegras before too—and their horses.

It was 2009. I had graduated from Columbia's journalism school, and after months of dialing every phone number I could dig up for the *New York Times* and pitching stories to anyone who picked up, I had rammed my way in as a prodigious freelancer. After a year of that, I got my first full-time job at the newspaper. That was when Carolyn Ryan, an editor of the Metro section, asked me to become the Nocturnalist—the correspondent for the *Times*' new nightlife column. I was selected in part because of the gall I'd shown wedging myself in the *Times*' door, she said—you need chutzpah to harangue celebrities too, after all. But in large part, I'm sure, it was because I had the youthful stamina to stay awake that late. As Nocturnalist, I covered 252 parties in the eighteen months of the column's life before I was drafted into a new role writing breaking news.

One fashion week, I hit twenty-five parties over the course of five days. Racing between the shindigs that week, I vividly remember slinging my stilettos in one hand over the photographer's shoulder as I vaulted onto the back of his motorcycle. "To the next party, and step on it!" I yelled. The pair of us almost tipped the bike over, laughing at the absurdity that was our job.

One February night in 2011, we revved up his Harley and zipped to the Waldorf Astoria Hotel, slaloming between the limousines crowding the Park Avenue entrance. It was the annual Viennese Opera Ball, a gala inside the encrusted upper-story ballroom of the storied hotel. Ostensibly it is a fundraiser for Vienna's arts, but it is also a debutante coming-out party. That night, a cadre of upper-crust debutantes would make their high society debut, each on the arm of a young cadet in the navy and scarlet of formal military dress. As I pulled up, the girls eddied on the pavement in front of the Waldorf in their white gowns.

Some wrestled their trains free from the backseats of limos or tugged up one another's elbow-length silk gloves, all exclaiming the way girls across the world greet each other always.

We buzzed the motorcycle past the gaggle and turned down East Fiftieth Street. At the back door of the hotel stood two more guests all in white: a pair of snow-white Standardbreds. That was where I first met Lori and John Allegra: they had supplied the carriage horses for the ball, shipping them in from their Allegra Farms in East Haddam, Connecticut. The horses, two others from their herd named Hayward and Zorro, were nonchalant on the pavement, tearing strands of hay from where it hung in nets from the side of their horse trailer. That night, they too would make their debut on the Waldorf ballroom floor.

For more than five decades, the centerpiece of the gala, more so even than the glittering debs and their escorts, had been an equine spectacle. It is made the more eye-popping by the fact that the equine guest stars perform not just amid the tables laid with bone china and crystal, and among the dowagers with diamonds draped down their décolletage, but because the horse show happens on the building's third floor.

Just before the debut, the horses were pried away from their hay bales. They stepped smoothly off the street and into the ground floor of the Waldorf, and without so much as a blink, they walked into a mammoth freight elevator.

Hayward and Zorro followed auspicious passengers into that elevator, including Franklin Delano Roosevelt. Roosevelt had taken great pains to hide his polio—and his need to use a wheelchair—from the public. To do so on visits to Manhattan in the 1940s, so the legend goes, FDR made use of a secret train station beneath the Waldorf, known as Track 61, according to news reports at the time. His private presidential locomotive would chug under the grand hotel, where he would then drive his armor-clad Pierce Arrow limousine into the giant elevator. President and car would be whisked upstairs, without his ever leaving his limo and exposing his secret.

As the elevator rumbled up toward the third floor, filled with two journalists, two handlers, and two horses, I had to repeat silently to myself that compared to an armored car, our weight was nothing. It climbed skyward, and the doors opened into a clattering prep kitchen. Unfazed, as if they were merely walking to their paddock, the horses snaked through the racks of cooking pots and pastry tables straight to their dressing room, in a manner of speaking. When they arrived under the heaving crystal chandelier in the Basildon Room, a farrier fitted their hooves—which were studded with grippy carbide for the icy roads in East Haddam, where the two normally did weddings—with rubber booties. Heaven forbid they scuff the polished ballroom floor!

The Standardbreds performed admirably in the ballroom, wobbling only imperceptibly under a spotlight that followed them around the space, trotting to the accompaniment of clinking crystal. The pair obligingly towed one of the Allegras' 1899 wicker carriages across the dance floor. In it, a ballerina and her partner performed a duet as the horses slalomed between the glitterati munching on poached salmon and beef Wellington.

The debs were less well mannered. "If *we're* in full white tie, and completely bedazzled," one escort named John Munson said to me peevishly, I later reported in the *Times*, "the horses should be too."

Those nightlife days are behind me, and at the time I took

that job, I worried that covering the fripperies of the world would pigeonhole me as a reporter, sidelining me into fluff stories. But by the end of my sojourn in the night, I felt quite differently.

In my final Nocturnalist column, a few months after the Waldorf ball, I wrote: "'Don't you get bored at all those parties?' is one of the questions asked most frequently of Nocturnalist, and one of the most surprising. Parties are never boring when viewed David Attenborough–style: as a chance to study people in the wild. People, it turns out, are the most fascinating creatures on earth."

Backstage at *Aida*, I was once again in that extraordinary, pinch-me position of observing a rarified nocturnal spectacle. At Nancy's signal, first Córdoba, then Nacho trotted out under the proscenium, leading the army home from war. The blue feathers arched over their brows as they pranced, bobbing almost in time to the music. Next, she gave a nod, and Billy and Bobby strode out of the wings in tandem, Radamès the warrior proud in his chariot behind them. John, in his toga, led his Fjords across the stage. He never stopped surreptitiously stroking their noses, centering the animals as the notes swelled through the opera house and the audience cheered.

As I watched the animals enter the stage, implacable, obedient, and brave, I might have revised my thought about *humans* being the most interesting creatures in the world; horses are at least are a very close second. The four horses entered the stage in the thrall not of the music but of the jobs set before them. Here they were, thrust into a scenario beyond any equine's imagination, a mind-boggling scene underscored with bassoon and viola, swirled with dancers and divas, ancient Egypt in the middle of Manhattan, and the horses were, as ever, whatever we humans asked them to be. They were stars.

From somewhere out in the red velvet seats, I could have sworn I heard a little girl squeal.

MASTER DAVID

No one else in my family likes animals—unless they come with a béarnaise sauce. But for a brief moment, my eldest brother had a racehorse.

My three brothers are much older than I am, the eldest by twenty-two years. In 2004, my oldest brother, Danny, a man for whom animals are an anathema, and horses in particular a terror, obtained the ultimate token of resplendent success: a racehorse named Master David.

Danny is a self-taught hedge fund wunderkind who eclipsed even my father's wildest imaginings for himself, retiring at age forty-five to a life of philanthropy and golfing around the world. Dad would crow about how Danny, through his canny understanding of financial markets and his own ingenuity, had catapulted himself into the winner's circle. For Dad, it was more than a boast: the ascendance of his children was armor that would protect us in case the world ever turned on our people again.

"Our father, I think he admired success, he liked the absoluteness of success," Danny said to me when I called him to talk about what on earth had possessed him to buy a racehorse. (Of course, he and I started out talking horses and ended up talking Holocaust, the trauma lurking behind every part of our lives.) "But in reality," Danny said, "that element of wealth would really protect you? His father had been relatively wealthy, and it didn't protect him. But our dad's hyperbolic assessments of achievement gave him some sense of identity."

That afternoon when I called Danny about Master David was one of the only telephone conversations my eldest brother and I have ever had. My two oldest brothers are from my father's first marriage. Dad was married to my mother, his second wife, for forty-five years. But still, I am the fourth child and only daughter born into a narrative of divorce unfolding long before I, the last child of my father's new marriage, was written. My brothers have nicknames for one another and blame their distance from me on passive factors like the more than two decades that divide me from the eldest. It's not true. The truth is that to them, I was an invader, my existence imbued with the heartbreak of their own family that fell apart. The story I entered into was already written. I didn't even lift a pen.

It is more painful than having no brothers, having brothers you don't have. As a little girl, I used to wonder at what offense I had caused, sitting for hours at four and five years old behind the swinging kitchen door in our apartment, wracking my mind to remember when I had done the bad thing that made them keep their distance. Of course, the answer was divorce, a wound I had not inflicted but came to embody for them. I didn't know that then and for long years of loneliness. Their family crumbled, and mine sprang up from the ashes of the life they were supposed to have with my father. I became the locus of their rage. I spent decades wondering what I had done to make them not want me. It took me a long time to realize it was being born.

Twenty-one thousand gangly, knock-kneed registered thoroughbred foals are born every year in the United States. They become the contenders bound for the country's one hundred or so racetracks, future Master Davids. And each one of them bears a coded tattoo on the interior of their upper lip.

Those tattoos are the brainchild of the Thoroughbred Racing Protective Bureau, an agency established in 1946 to clean up a shifty sport. To be clear, "Protective" does not mean animal cops bent on stopping beatings, the running of lame horses off their feet, or the callous discarding of talentless creatures. That still happens, in this and all other horse sports, and is policed to varying degrees by other entities. The bureau, however, was set up to protect the sport's integrity—that is, to protect its pockets.

Modeled after the Federal Bureau of Investigation (more than modeled; its ranks were in fact stacked with ex-FBI agents), the agency developed the tattoo identification system. Tattoos are registered in a database maintained by the Jockey Club. (As of 2017, baby thoroughbreds must be also injected with a rice-sized microchip in order to be officially registered with the Jockey Club, the breed registry. It is lodged in the animal's nuchal ligament, the sinew that traverses the topline of their neck.) Tattoos consist of a letter that corresponds with year of birth—A for 1997, B for 1998, and so on—followed by a four- or five-digit serial number, kept on file by the Jockey Club. Although as of 2020 tattoos will be phased out, inspectors for decades have roamed tracks, tattooing and microchipping horses to verify who the animals in fact are. The Protective Bureau's goal is to prevent and suss out horsey doppelgangers.

Ringers.

The thing about racehorses is that the breed is usually brown, reddish, or gray. A few have distinguishing white markings, a dribble splattered like Jackson Pollack painted their hides. One or two or three or four legs may be dressed in white fur, bobby-socked or elbow-high like Holly Golightly. Their tapered faces may bear a creamy half-moon, a splash or a starburst, or a drib of whitewash down their nose. These white markings are known, for some reason, as "chrome." But chrome is nothing that can't be erased with a bottle of, say, human hair dye.

Racing has been plagued by the ringer, as in horses who are a dead ringer for one another, with perhaps a little extra help from some bootblack or hair dye. Prior to the bureau's authentication procedures, it was simple to swap one horse for another. Why would equine identity theft be a problem? Race-horses go up against one another in heats where the animals running out of the starting gate are matched by skill. It is much like how boxers are divided into bouts by ability and heft, so that a welterweight doesn't get pulped by a heavyweight. If you could somehow substitute, unnoticed, a heavyweight for a wel-terweight and have a bruiser spar with the ninnies, you'd have a guaranteed winner. You'd have a ringer.

"You've gotta come up with a bum," said Paul Berube, the

former president of the Protective Bureau, who retired in 2005, describing how the schemes went down. "Horses are not all equal when they race," he said over the phone from his home in Kemblesville at the semirural, southernmost edge of Pennsylvania. "You have horses that compete at the highest level—the American Pharoahs of the world, the very cream of the crop. Then you have a long trail of different class levels and levels of abilities, all the way to what we might call down to the bottom." He stopped for a moment in the middle of our interview, as if deciding whether to insult the animals that for his five decades with the bureau had been the totality of his world. "The cheap horses."

The bum of the century was a horse called Lebon, foaled in Uruguay and imported to America in 1977 by Dr. Mark Gerard, a track veterinarian who had cared for, among others, the Triple Crown champ Secretariat himself. Lebon was an unspecial bay with a sprint that a *New York Times* article on him would later say "couldn't beat a fat man from Gimbels to Macy's," two city department stores nearly side by side, and a blot of white between his eyes. On the turboprop plane during the horse's importation from South America, beside Lebon there was a starlet: the top-performing Uruguayan thoroughbred, named Cinzano. A year younger than Lebon, Cinzano was a bay horse glossy with power—and a blot of white between his eyes.

The day after Cinzano and Lebon arrived at Dr. Gerard's Muttontown, Long Island, farm, Dr. Gerard was dining with friends when he got word of a terrible stable accident: Cinzano had fractured his skull. By the time dinner was over, Cinzano had been euthanized.

Just a few months after the death of his champion stablemate, Cinzano, Lebon went through a stunning transformation. On a September day at Belmont Park racetrack in Elmont, New York, Lebon the loser outran the other horses. He won that day at 57-to-1 odds, pulling in nearly $80,000 for his owner on a minuscule $1,300 wager. Flashbulbs popped in the winner's circle, and even in that pre-Internet day, word of Lebon's unlikely upset spread all the way back to Uruguay. Over five thousand miles away, a local Uruguayan track journalist did a double take at the picture of Lebon, the bay horse with the blot of white between his eyes.

Cinzano.

The international lines lit up, from Uruguay right back to the United States, where a certain brown colt was apparently running races from beyond the grave.

"The record reveals a factual scenario that might have been authored jointly by an Alfred Hitchcock and a Damon Runyon," the New York Court of Appeals wrote in its ruling in 1980, after Dr. Gerard and others had been found guilty of fraud, the appeals panel ruling to uphold his conviction.

That stable accident in Muttontown had been no accident. It appears that Dr. Gerard had arranged a hit on the real Lebon that night in the barn but pretended that the tragedy had befallen Cinzano. That allowed Dr. Gerard to switch the horses' identities. He ran the star colt Cinzano as the nag Lebon in lower heats, where Cinzano would beat all the animals in his path by furlongs.

Dr. Gerard's scheme played out as far as it did because the foreign-born horses were raced without the same tattooing stipulations as American thoroughbreds. Nothing was inked under their lips when they arrived in America. Ever since the Protective Bureau was set up in the 1940s, horses in the United States must carry the identifying numbers if they are to race. Import the blank-lipped champions like Cinzano from Uruguay, or Argentina, or where have you, and you had a window to tap new identities into the mucous membranes of their mouths.

"They found a loophole; they found a way to beat the system," said Berube, who knew of the case as lore but had not been involved in it. "It's like anything else: you see it everywhere. What are people doing today with technology? They are hacking it and evading it and trying to turn it to their advantage—it's human nature." Here were hacked horses. Berube has spent his career in the sport outwitting and uncovering the hackers. At stake, he says, is much more than a payout at the betting window; it's the sport of racing itself. Yearly, horse racing is a $15.6 billion chunk of the $122 billion equine industry. The horse industry is responsible for the creation of about 989,000 jobs, from trainers to breezers to grooms to workers who shovel refuse, according to the most recent study commissioned by the equine industry's American Horse Council Foundation.

And then there's a sportsman's pride. "If it were just a rigged game like marked cards or loaded dice, who's going to want to play in something like that?" Berube asked. "Only your worst degenerates."

The internecine nature of Dr. Gerard's switch took years to unscramble. Confirming the fraud required forensic dives into equine dental records and investigators sent to the racing barns of Uruguay to bring back vials filled with Lebon's relatives' blood. Dr. Gerard was ultimately found guilty of misdemeanor fraud; he served less than a year in jail. He died in 2011. In his obituary in the *New York Times*, his sister confided he had left behind a handwritten note. Please scatter my ashes, it read, "where happy horses graze."

My oldest brother's two-year-long foray into horse racing was bizarrely successful. He owned a quarter stake in two horses, one of which, a mare named Buy the Sport, won the first race he started her in as owner, at 40-to-1 odds at Aqueduct Racetrack. He ducked into the winner's circle in formal attire, having popped by as a lark on the way back to Manhattan from a bar mitzvah in Queens. He was hooked.

Master David did even better, making it all the way to the Kentucky Derby in 2004, where my brother followed him on a private jet to Louisville. My brother had caught what's known as Derby Fever: an unshakable conviction that the twenty other horses hauled to the Bluegrass State for the meet—pumped to bursting with millions of dollars of meticulous training, juiced up with every mysterious substance in existence that won't show up on a tox screen, twenty muscle-rippled fruits of generations of calculated, selective, obsessional breeding—are in fact a bunch of nags. The Fever is also known as insanity.

My brother walked a little ways off from his gelding as the horse danced toward the starting gate in a drenching rain. A few minutes later, Danny's Derby Fever broke: Master David came in twelfth.

My brother left the sport soon after; he didn't like horses anyway, and in the owners' boxes from Santa Anita, California, to the stables where he had sloughed loafered feet to watch his quarter of an animal being breezed, or exercised, he felt he was not alone in the sentiment. "It's a bunch of guys trying to express their own testosterone by being identified with something that is beating something else. I'm not into that," Danny told me. "Or horses, really."

BENEDICTION

Horses became my siblings, when I had three but none at all. They were my playmates when I was young and my confidants. In the barn, I was grateful to be in the company of creatures who had nowhere else to be but by my side. I sat whole days cross-legged in the hay in their stalls as they gently lipped for strands around my legs. There, burrowed in the straw, I had the conversations with horses I longed to have with my family.

Horses talk, but not with whinnies like the Flickas, Mr. Eds, and The Pies of TV and movies, who seem to be perpetually bugling neighs full of meaning. The idea that horses would communicate with their clumsy, thick lips is deeply anthropomorphic. Yes, their mouths are perfectly muscled and tactile, but for ripping roughage and foraging through briar and snow. Horses are not built like us, they don't think like us, and they certainly don't speak like us. They converse mostly through body language, and their most visible correspondence is a sign language of ears.

Horses can flip and flick their ears 180 degrees. The orifices are controlled by ten muscles, as compared to the just three muscles tugging on the human ear. Horse ears are radar detectors, but they also function as mood rings. Just as lovers learn each other's emotional cues, riders learn to read their horse's expression, and woe to one who doesn't listen when a horse's ears speak. Ears tipped back in the direction of the haunches express displeasure and irritation. They also may mean the horse is just ticked off or mildly bored. And it's up to the human to know which. When they tip back farther, lay flat back, pinned to the skull, that's fury, the kind that can be underscored with a follow-up bite in case you didn't get the point. Ears akimbo, and a horse is daydreaming, thinking nothing much at all or maybe everything.

Pricked forward is a horse's smile.

It's always mildly startling when I think about the mechanics of the entire interaction between horse and rider—that the means between which our two species communicate while riding boils down to a conversation between bodies. "Whoa" and "giddyup" are the rudimentary commands for stop and go, but actual riding is a silent dialogue. Indeed, it makes sense that we don't for the most part verbally instruct horses while riding. Horses don't speak to one another; they don't chitchat. If you speak to your horse, you're not speaking in horse.

In the herd, horses school across fields like minnows. They turn and wheel like larks, guiding each other with shoulders and flanks, ear and eye, ripple of muscle and stomp of hoof. All are participants in an inaudible dialogue that nonetheless instructs them just as surely as troops would obey a spoken command.

"I've tried to stop calling it a language," Monty Roberts, the legendary horse trainer and bestselling author, known as "the man who listens to horses," said to me. "We think way too much about words and alphabets and stuff like that, because it's what we do with ourselves that gets through to the other human being we are communicating with," he added. "That's not horses."

Roberts, who is in his eighties, spoke to me from Flag Is Up Farms, his ranch in Solvang, California, which he's run since 1966. I had called him because I wanted to learn what he had heard in a lifetime of listening to horses. On his land in the Santa Ynez Valley, Roberts has been sculpting his understanding of horses into a discipline he teaches around the world. He calls the technique Join Up (and forgive me while I oversimplify the craft he has honed for over seven decades to absurdity), which asks the horse to work with the rider as a member of its herd rather than as its master.

There was cowboy grit in his voice, but also the homey cadence of Mr. Rogers. I felt immediately a friend, charmed by the same easy conviviality that convinces feral horses he's one of their own. Roberts had just returned from a morning session in a round pen, where the animal performs freely in a small enclosure, with a four-year-old Akhal-Teke with a brutal kick, and his voice glowed with a sense of accomplishment. At last he had met the tricky horse eye-to-eye. He described it as the most difficult he had ever worked with in his life. I rocked back on my heels. That's something of an achievement for a horse when Roberts's life's work has taken him to fix problem animals from Buckingham Palace in London to the Royal Randwick Racecourse in Sydney to the Spanische Hofreitschule in Vienna, where since 1572 they've trained Lipizzaner stallions to do arabesques in the air, and every animal in between.

"This took twenty-four days," Roberts said, clearing the valley dust from his throat. "I usually can get there in twenty-four minutes."

"How'd the Teke get so bad?" I asked him. "He's only four years old." That seemed to me a short time to make a hardened criminal.

Roberts danced around the answer for a moment, explaining the Akhal-Teke breed's desert wiring: that molten blood that makes them keen showmen but so sensitive that the concentric circles of a currycomb on their hide can be overstimulation. Then the truth: "He's had time to reap the harvests of the mistakes that human beings make," Roberts said. Of course, the colt's owners had insisted to Roberts they had no part in turning their baby horse into a creature that took the legend himself aback. Roberts saw otherwise in the Teke himself.

"You can believe 52 percent of whatever a human being tells you, but you can believe 100 percent of whatever a horse tells you," Roberts said, laughing softly. "They had a difference of opinion."

As Roberts strode over the pastures and through the pens of Flag Is Up in recent years, he had a little shadow at his side. It was a West Coast mule deer he found still wet with amniotic fluid in a stand of grass a few months before our chat. It was an unusual triplet fawn, a runt with little chance of surviving. Rearing the fawn changed Roberts's understanding of how horses communicate, his theories capable of evolving even now, after his decades of study. Roberts views deer as an equine exemplar; to him, they are the raw, wild, prey creature hyperattuned to its world for sheer survival that horses once were—before domestication bred the edge off. The Cervidae are Equidae at their most elemental.

The deer's name is Benediction.

"I worked with a lady in England named Elizabeth; she's the best namer of horses I've ever met," Roberts said. "So I emailed her and asked what I would name him. She emailed right back. She didn't say 'I suggest' or 'I think.' " She said, "His name is Benediction." I wondered if his eighty-plus years were showing and where the segue about naming this little slip of a mule deer was going. "So I called her up and I said, 'Your Majesty, he is a little humorous deer; "Benediction" is such a big name.' And she said: 'His name is Benediction.'

"Queen Elizabeth can sure name a horse," Roberts said as I snorted my shock. He giggled with satisfaction. "And a deer."

"What a horse does under compulsion he does blindly," wrote Xenophon, that ancient Greek horse whisperer. "And his performance is no more beautiful than would be that of a ballet-dancer taught by whip and goad. The performances of horse or man so treated would seem to be displays of clumsy gestures rather than of grace and beauty. What we need is that the horse should of his own accord exhibit his finest airs and paces."

A pupil of Socrates, Xenophon was a cavalry master who died in 354 BCE, a warrior who survived the battles of Sparta. His treatise, *On Horsemanship*, is one of the earliest surviving works on the art of the equine—the selecting, gentling, and toughening of a warhorse. "The majesty of men themselves is best discovered in the graceful handling of such animals," Xenophon wrote. (I'd like to add "of women" too.)

Taming a horse, gentling it, or, crudely, breaking it, involves messaging more than anything. Horses are so huge that no amount of force a human could use can truly push them around. Horses are rideable at all, in a way that, say, lions and tigers are not, in large part, I believe, because equines are prey animals, bound to the herd. Their mental mainframe comes preprogrammed with the inclination to submit to a leader, an alpha, for safety and guidance. The more individualistic thinking of a predator, the cunning that keeps it ahead of the pack and makes it the first to sink its teeth into the quarry on the savanna, can get a lone horse murdered. Horses are genetically inclined to accept a boss. And me.

The phrase "broke to ride" used to mean the animal's independence was knocked from it, a brutal process in which its spirit was broken and the shell left behind submitted to human will in defeat. In modern equitation, other terms have become popular, like "backing" a horse—getting it to accept a rider on its back—or "starting" the animal. At its best, the process today is something more like recalibration, convincing a horse to its core that you run the show. A horse may be twelve hundred pounds, but successful training convinces a horse that size doesn't matter; *you're* the herd leader. Done well, submission is rebranded as alliance; the mount and rider, a herd of two.

Roberts has had a lot to do with that shift. Growing up in the 1940s on his family's ranch in Salinas, California, Roberts saw that violent methodology wreaked on horses—and his own young body. "I had seventy-two broken bones before I was twelve years of age," he told me, "from a father that wanted me to keep my mouth shut about everything I saw." When his first book came out in the mid-1990s, the bestselling *The Man Who Listens to Horses*, the radical departure from the prevailing discipline (dominance and cruelty) earned him enemies. There were even death threats, according to Roberts, leading to the arrest of at least one person. "You're telling them that everything that they've done in their life is wrong," Roberts said, trying to make sense of why he was so hated for saying simply, *be gentle.* "Not just that, you're telling them that their daddy's methods, and their daddy's daddy's methods, were wrong.

"A good trainer can make a horse do anything they want it to do," Roberts said. "A great trainer can cause the horse to want to do it."

Through Benediction the fawn, Roberts honed further his understanding of how horses communicate. "The ears, the eye, the neck, the lowering of the head, the licking, and chewing the tongue, all of the appendages or parts of the anatomy of Equus and Cervidae are put to work to let the others know what the reading of the situation is," he said. And all that anxious study by the animal is to answer one question, he added: "Is there danger?"

Roberts witnessed that intra-equine communication as a thirteen-year-old boy, when he was sent to Nevada to round up mustangs for a rodeo race. Observing the intricate herd dynamics, the solicitousness of the studs, the defiance of the mares, the submission of the spindly foals who skittered among them was when he first realized horses communicate nonverbally in a clear, delineated system of gestures. He called that language, that dance, "Equus," naturally. In deer, he sees Equus amplified.

"The major driving force of every gesture that is made in Equus the language," he told me, "is to find a safe place to be."

* * *

For a year of my life, Equus was my language. Because on Thanksgiving Day of 2010, I became prey.

Dawn had not yet broken in my apartment in the West Village. I slept fitfully that night, zapped with nervous excitement for the morning: my first time covering the Macy's Thanksgiving Day Parade for the *Times*. Bev and I had always squished among the spectators when I was small, craning up at the giant balloons of Snoopy and Popeye floating down Central Park West. Later that day, notebook in hand, I was to walk the storied route with them, underneath the shadows of my childhood icons.

In my dark bedroom, through my closed lids, I felt a different shadow.

Suddenly, I was awake—or was I dreaming?—and I was fighting, squalling, kicking my legs as fists descended into me over and over again. I was screaming, but my voice betrayed me, and my throat made no sound. As I was punched and beaten, deep inside me I heard a truth: "This is not a fight you can win, Sarah. Find another way." I stopped fighting in an instant. I lay still.

The man in my bedroom smelled like smoke, amplified, like a thousand stubbed-out Marlboros. The intruder wanted cash, jewels, electronics, stuff, he told me, and instructed me to lay prone as he ransacked my home, his sour body pulled into a dark corner so I wouldn't see his face. I would survive, I decided that moment, by being the most helpful victim of all time. Confined to my bed, as he rooted around I told him how to find everything of any value. I chided him to get a pen and paper from the kitchen, so he would be sure not to forget my ATM pin code for whenever he went to the bank to wipe out my savings. When he discovered the only valuables in my tiny apartment were a single laptop and a fistful of costume jewelry, I cracked the New Yorkiest of jokes to appease him: "You know how Manhattan real estate is, we spend all our money on rent!"

The stranger had climbed through my second-story window, New York City Police Department detectives would later tell me. He left through my front door. "I swear on my son's life, I won't hurt you," he said when he was finished robbing me, and the lock clicked closed. That was when I realized he already had.

Alone again, I lay in a pool of my blood; it poured from a four-inch wound in my leg. The man had stabbed me with a box cutter in that first brutal struggle. I hadn't even noticed. Adrenaline had stilled my wild mind and made me into the world's most obedient prey. It was my compliance that detectives later told me had saved my life by keeping a jittery man with a weapon from using it to silence me. And it had made me insensate to pain. As I watched my blood coursing from the wound, I was terrified, but also elated—my pulse, pooling on the floor, was proof I was somehow still alive.

It took police just a day to catch the man, who confessed it was a crime of opportunity, that he broke open my life simply because my window was open a sliver. "Thank you for this holiday gift," the forensic crime-scene analyst said after I was stitched up at the hospital, while she hunted his fingerprints in my apartment on Morton Street. "In cases like this, I'm usually dusting for these prints off the dead body."

The man was sentenced to seventeen years in prison, but as I limped around the city and tried to recover, I was also trapped. New York's cacophony was my childhood lullaby, but suddenly the city was loud, so loud. For a year, I became the Cervidae, my body suddenly hypervigilant to every sound. Like a deer, my body was listening for him, for a city of him, box cutters bared, and couldn't stop. Air conditioners whirring on the sides of brownstones were buzzsaws, made scarier because their white noise blocked me from hearing what else might approach. The grind and hustle of the metropolis, the backing track to my urban upbringing, became a predatory screech and a villainous roar.

If the language of Equus is to find a safe place to be, the language of my once-beloved city shouted to me at the top of its lungs: nowhere is safe anymore!

In horses, hyperattunement to their environment and to their peers keeps them alive. But it's also why horses can "hear" us humans and respond to our bodies, like the pressure of our heel that says "trot on," or the rebalancing of our seat bones that asks our mount to steady. "Everything they do—reading your intention through cortisol levels and pulse rates and adrenaline levels—relates to that," Roberts told me many years after my attack. "Reading that from afar is their way to survive, and they do it better than any human being ever would," he continued. "Reading it close up—a horse can feel the artery in your inner thigh pulse through the saddle—is why they can be ridden."

During those dark, loud days of my life, safety was among horses, among quiet beings who talk only with their muscle and breath. How did I learn to trust the world again? It was the same way a foal learns to stand—in that it doesn't actually learn. It just does. It gets up, falls down, gets up, carries on, because it must, because that is living. I stumbled forward, and like the foal, I fell, and at age twenty-seven, moved back for months into my family's apartment. Once again, I crept into my parents' room each night as I had as a child.

Benediction means a blessing. And as I kept stumbling on, I felt less afraid and more keenly the blessing of being here, alive, still, even if it was just to stagger forward. On Thanksgiving Day 2011, I walked down Central Park West underneath Snoopy's big helium belly, notebook in hand. There in the middle of the parade, the city was no longer so loud. Now, I work almost every Thanksgiving, to cover the day I almost missed, forever.

Unlike a fawn or a foal, I realized that whether to live as prey was a choice I could make, not one made for me by a stranger in the dark.

WILLOW

"No one ever came to grief, except honorable grief, through riding horses," Winston Churchill wrote in his 1930 memoir, *My Early Life*. "No hour of life is lost that is spent in the saddle. Young men have often been ruined through owning horses, or through backing horses, but never through riding them. . . . Unless of course they break their necks, which, taken at a gallop, is a very good way to die."

To fall off is to ride. From the dirt looking up at Guernsey, to the hospital beds and whirring CAT scan tubes into which other horses have sent me, I've always known it is an immutable rule of riding—and that pain is as much a product of the sport as its joy. To this day, my right foot drags slightly, wearing down my heel on that side at twice the rate of my other foot, the lopsidedness of broken bones I refused to let heal, too desperate to get back in the tack. My back quivers with electrical agony at all times. On a pain scale of one to ten, my daily measurement is a six; perpetual pain is part of my life. But so are horses. At points after I've fallen off an animal or done nothing more dramatic than make my bed, I'm petrified by pain that lasts for days. It renders me immobile except for embarrassing tears of self-pity; somehow, no matter how stiff I am, those flow just fine.

To get back on is to live.

Willow was the seat of my highest equestrian glory and my lowest low. She was as slim as a saluki dog, those tasseled Persian greyhounds so slight I once exclaimed out loud at one flouncing down Madison Avenue because I thought it was actually two-dimensional. That Willow could, if she so chose, slip between a door and its jamb was not her fault but the fault of her race: she was 100 percent pure thoroughbred.

The thoroughbred, in all its magnificence, is an Arabian enlarged, elongated, lent mass and brawn by interbreeding way back when with heavier English horses. It is a breed hybridized for centuries from Arabs, lithe as caracals and sensitive as hothouse orchids, into creatures dense of bone and born for flight. The creature that emerged, codified in the middle of the eighteenth century in the burgeoning business of jockey club logs and studbooks, is a tank engine with seven-league boots. The horse that was produced is so iconic that when you think of a horse today, it's a thoroughbred you see behind your lids.

Willow was a dapple gray and her neck was a swan's. She held her plain head high above her shoulders in the manner of an imperious ballet dancer looking down at the affront of being offered dessert—so much so that her neck inverted in an S-curve. It's called "ewe-necked," and it's not a desirable trait—not that I was an educated enough horsewoman at sixteen years old to know that then. Camel-like, I would have thought, if I had been capable of looking at her with anything other than abject reverence and adoration so complete I would stand before her in the barn aisle in a stupor.

It was 1999, and I had shot up and outgrown my beloved, stumpy Amigo. We traded him to the owners of East End Stables, and he lived out his days as a walk-trot horse, scared of plastic bags until his last breath at age thirty-four. Plus his trade-in value (not much), my parents were able to put enough together to purchase a horse, albeit a discount one: Willow was on the bargain rack because she was gaunt and sallow through no fault of her own. Her ewe-neck was from poor conditioning, I would later learn. Her ribs xylophoned along her sides. Instead of powerful haunches, that muscled stuff of legend, the oil-painted magnificence of glossy Stubbs portraits that hang in dark rooms above mustachioed men in leather chairs, her rump was a peak of pelvic bone.

There is a story behind every horse one meets, and the silence of the animal itself is deafening. Where did the white slashes of fur regrown in the wrong color on the bay's withers come from, those memories of scars that spackle the hide of almost every animal? Who let the saddle fit so poorly it dug into that tented spot where your shoulder blades meet? Who kept riding as blood fell? Why this dent in the flesh of the neck? The puckered skin across a cannon bone? Running hands over an animal, you can feel the history in its hide. It is an exercise of answerless questions. When did that happen? And: Who did that? And: Who hurt you? And always: I'm so sorry.

Willow was not abused, but she had been left fallow in a field. Her owner, a woman who lived not far from East End, was pregnant. With no time for her horse, I was told she had turned Willow out to pasture. People without exercise get fat. Without exercise, horses devolve. Their flesh hangs, their joints loosen. Unmoored by muscle, necks flatten and hollow into the cheeks of old crones. Unexercised, a horse is unwell. Willow was unwell. And she was mine to heal.

Most horses, with the exception of a specially bred few, have four gaits: the walk, the trot, the canter, and the gallop. (The Icelandic Horse can have up to six, including a syncopation-silencing swagger called a *tolt*, which gives a feeling of sudden hooflessness, like you're riding on a hovercraft.) Walk and trot are parallels to the walk or jog of a human, just with more legs in the mix.

The canter is something different entirely. One side of the body leads, like a dancer's sashay beneath the proscenium. (Or really anywhere, but I can't help but think of the stage whenever a horse moves with power. Their movement, like waves curling in from the far sea, seems to me solely created to be watched.) For balance in ring riding, when you change directions at a canter, your horse must swap leading legs or, as they say, change leads.

Called a "flying lead change," it is a deceptively simple move. It is slowly impressed into young horses in training, over years, and whether or not a fully grown one has grasped the concept can dock the creature's value as completely as a bowed tendon. It is a clockwork of timing and brawn, requiring the horse to launch itself in the air for what is, in essence, a full-body gear shift.

But it's far from dangerous, so no one was more shocked than I with what happened on my third ride on Willow. On a sunny June afternoon, we cantered across the ring, and I asked her to execute a flying lead change. Instead, she caught a toe and torpedoed to the ground.

I don't have any memory of that. What I remember is her long, thin face when she crept back over to me, where I was prone on the ground, and hung it over my body. It was a mask of blood, too bright on her white fur. It poured down between her eyes, over the place on her brow where, for the few weeks I had owned her, I had stroked the silver dollar of pink skin hidden underneath her down. All gray horses are born dark, with black skin underneath, and they white out with age. The little patch of pink beneath her white fur was evidence she had been dotted with a hoary star when she was a jet-black foal, and I would stare and wish I had known her then and so could have always loved her.

Behind her was the white wooden fencing encircling the ring. A four-inch-round pole lay split in half. Willow had broken it with her forehead. Blood also came from her lips, smearing her mouth darkly. It mixed with dirt; she had smashed her face into the ring dust. Standing there, she ground a dark paste between her molars. In the sand ring, I heard the grumble of her jaws grinding—the sound of broken teeth. Such an injury could be a potential death sentence for a horse, born of grassland to grip and rip and eat.

I jumped up, to cracks and pops that I later learned were inside my body, but at that moment felt external, fissures splitting my entire world. I had to save her teeth! I swiped off her bridle, removing her bright steel bit embedded with the sweetish copper beads that she liked to worry with her tongue, that would settle her taut thoroughbred mind. I braced for a cascade of her incisors to follow.

Instead, out came a mouthful of dirt; it was the grit she'd scooped under her tongue as her nose plowed into the ground that was making the sound. She'd smacked her head and would later be diagnosed with a concussion that left her meek and wobbly, but she'd recover. As I took the bridle off, I realized that aside from that, she'd only bitten her lip.

I sunk to the ground then, when I knew she was for the most part okay, adrenaline for my horse giving way to my own pain.

At Southampton Hospital a few hours later, I learned I had broken three of my vertebrae, a physical blow that gave way to a deeper one: they told me I could no longer ride.

I stood at the podium in the Riverside Memorial Chapel on Manhattan's Upper West Side on an afternoon in 2014 before a crowd of five hundred at my father's funeral. It was a room packed with the characters my father had picked up at the opera, at museums, and at concerts. The room bore the fruit of his lifelong habit of bringing interesting strangers home for dinner. My father was beside me, shut up in a plain pine coffin as is Jewish tradition, about to be overnighted to a hilltop overlooking Jerusalem. He is buried in Israel. Atop his grave is a marble slab. The Hebrew thanks Ludwig Selig, the artisan who crafted Dad's false baptismal certificate and was murdered for saving all of our lives.

At his funeral, I spoke of horses. I told of the afternoon, when I was seventeen years old, six weeks to the day since my fall off Willow and against doctor's advice, I got back on. It was the Hampton Classic, one of the most elite horse shows in America. Showing at the Hampton Classic is all that generations of pigtailed pony riders like I once was crave. The show has been held in some form since the 1900s and takes place in

Bridgehampton, a gilt-edged town. The sixty-five acres of the show grounds are used for just a single week the entire year. For those seven days, under a circus of blue-and-white-striped tents are stabled world-class horses, the crème-de-la-crème of show jumping shipped in from around the world. Olympians use the Classic as a proving ground for future horses they will take to the Olympic Games; Hamptonites, prowling the show grounds with the golden "H" of Hermès belt buckles at every waist, use it as a proving ground for their high net worth.

This was the world where Jackie Kennedy, when she was nine years old, competed her horse Stepaside. At the Classic, even the toddlers towed by lead lines on ponies the size of a St. Bernard do so in Ralph Lauren. During its six days of intense competition, coming home with a prize is typically done on a six-figure horse so well broke he could complete the event on his own, no rider necessary. For horse girls in the final moments of summer vacation, the Classic is the finale of a season's worth of equine devotion.

Sidelined from riding that summer, I wrote instead. The *East Hampton Star*, the town paper since 1885, run out of a shingle-covered building with a gambrel roof on Main Street, gave me my first reporting gig. The *Star* is an intentionally small-town broadsheet, the kind where the editorial board fires off screeds about the village council's decision to swap roadside welcome signs to the hamlet of "*The* Springs" with just "Springs." ("The *Star* supports this decision, having been in the no-'the' camp since the 1960s," the editorial board opined.) So the horse world in its minutiae was big news, and in my convalescence, I was made teenage editor of the *Star*'s Hampton Classic special edition.

A summer spent under a currycomb, munching lush grass and being ridden for me during my convalescence by my barn friends, had returned Willow to the racehorse she once was before she began her current, second, career as a jumper. She was full of flesh and muscle and, at last, health. That made one of us. Four days before the competition, my body screaming, I put down my pen and got back on.

The morning of the Hampton Classic, Willow and I tried our hardest. Memory can manufacture overidealized versions of the past, but she and I really did soar over manicured hedges across an emerald lawn. If I had written about it for the *Star*, I would have said that the land unfurled beneath us, a perfect tapestry of green, the grass uncrimped by hoof or heel 358 days a year. But I was off the newspaper's clock, wholly present. For once, I entered the show ring without my typical amateur fear. I'd already broken my back. What more was there to lose? was my *insane* rationale. The late-summer air was quiet and still even as we galloped, vaulted, turned, galloped again at the next obstacle. Willow was a ballerina, throwing arabesques over the rolltops and birch rails, and I was her dance partner, matching her every move with my own.

We stepped out of the ring and I petted her damp neck. I didn't savor a moment of it. I didn't even realize what we had done, so convinced I was that someone like me couldn't possibly hold my own against the most elite horses and riders in the country. And there were more than sixty of them in my heat. Sure, I was there, but I did not belong. This was Jackie Kennedy's world. And in comparison, who was I? Sarah Nir? Sarah Gruenfeld? Did I even know? I dismounted after my round of jumps, bathed Willow with a garden hose in the parking lot, and put her away in the trailer with many lumps of sugar. I went to find some funnel cake.

But my dad stayed alone at the ringside for hours, leaning over the privet and watching every round. He knew nothing of horses, nothing of how I stacked up, but he waited under the late-summer sun, his bald head crisping, to hear the results. At day's end, over the loudspeaker came the announcement, and one by one the riders strode in on their mounts to the winner's circle to collect their grosgrain ribbons as their names were called.

And there, near the front of the equine processional, shuffled an elderly Polish man. There were sixty riders, and I had won second place. I was nowhere to be found. So my father strode into the winner's circle to collect the red ribbon. He turned to the judge, raised up his arms, and uttered the immortal words: "I defeated Hitler!"

For years after, Dad would tell the tale that he called, "From Pepper in the Ass to the Hampton Classic," reveling in his caustic framing: the interloping Jew besting the Aryans at their own game. I had wished our horsey history had been genteel, but my dad's vision of us as scrappy subversives had its appeal. Unlike me, Dad did not wish to assimilate into some rarified other. He didn't feel like an outsider; he felt like a champion. His view of our life—shoehorned into a world where we were too ethnic, too kosher, too culturally unmoored from its WASPy mores— was of a great and lifelong victory.

At the funeral, I told that story to the pine box beside me, my last words to my father himself. *Ritorna vincitor.* That red ribbon was the spoil of my father's personal war.

SNOWMAN

The real reason I left the ring at the Hampton Classic that day had nothing to do with a craving for funnel cake. It was because of a man named Andre. He was the owner of East End Stables, where I kept Amigo and then Willow. Andre is the son of the storied Harry deLeyer, a Dutchman known as the Galloping Grandfather for continuing his winning ways well past his fifties. Gray-haired and stooped in the saddle, the elder deLeyer would toss his velvet helmet in the air and canter to catch it every time he beat the buzzer in a jump-off round. But Harry's big claim to fame is the well-worn saw of Snowman, the show jumping horse he bought for the price of its flesh at an auction in 1956: eighty dollars.

Snowman was a coarsely hewn creature, but his mind and soul were deeply refined; he was a children's pet as well as an unbeatable jumper. Harry liked to show off Snowman's scope and demeanor by making him vault other horses the trainer forced to stand between jump standards. Not long after he bought the gray, his children's lesson horse won the Diamond Jubilee under the lights at Madison Square Garden, clearing jumps higher than the animal's head. Snowman went on television and once obligingly let Carson clamber atop him with a stepladder and plop facing backward on his white back. Some called him the "Cinderella Horse." The name fits because to me, it has about as much truth as a fairy tale.

The deLeyers espoused a riding discipline that now fills me with regret, learned, as anyone who has trained with the extended family (there are eight children, several of whom are equestrians) knows, from Harry. They seemed to possess an antiquated sense of the animal as a beast to be subdued—that conception of a horse that Monty Roberts has spent his life fighting—rather than trained and encouraged to perform at will.

They're not alone, and others are far, far worse. And it's not over. On a glittering day in Bridgehampton in 2019, I watched the world's elite riders compete for a $700,000 pot atop million-dollar show jumping horses. I couldn't even bring myself to applaud: One rider had previously admitted to smashing a horse's leg to collect an insurance payment. Another, two years before, had been temporarily banned for burning his horse's legs with a caustic substance to make them so sensitive it would do anything not to touch the jump rails. And another, an Olympian whose father was convicted as part of a scheme in the nineties to electrocute horses for insurance money, was himself barred temporarily from the sport as a younger man for embedding plastic shards in his mount's shins.

None of them won. The winner of the 2019 Hampton Classic Grand Prix was a Canadian Olympic rider who has been suspended at least twice from the sport, once for a horse who tested positive for human antidepressants and a previous time for cocaine.

At East End, my mind and soul were burrowed in the manes of the beautiful school horses who were my charges to comb and feed while I worked at the stables. Blue, Taxi, Ghost, Noid, Strawberry, Gingersnap, C.C., and my adored Amigo and, later, Willow. I didn't question Andre and his wife, Christine. It took me until adolescence to realize there were other ways to train horses.

Sometimes I can't sleep thinking about how I followed orders to remove the water buckets from the show horses' stalls the night before a horse show so they would be "quiet" and subdued on competition day. Twenty-four hours of competition without a sip of water? Blazing summer days in which my horse would try her heart out for my entertainment, and I forbade her to slake her thirst? And I perpetrated such a thing myself?

From her racetrack past, before Willow was retrained into a show jumping horse, she was lopsided. Many thoroughbreds are, from a lifetime of loping in races that always run counterclockwise. One afternoon, I was pulled off Willow as an East End trainer swung on, determined to force the crookedness out of her. I watched at the fence line as Willow was whipped in pinwheels, the bit cranked so that the horse's head almost touched the woman's thigh. For the next forty-five minutes, the mare was spun in circles until sweat foamed on Willow's neck and the lather slid down to her hooves. "Fucking crooked horse," the trainer said before dismounting and tossing the reins to me. It makes me ache to realize I was part of it by not speaking up.

But I stayed at East End for eight years, despite my misgivings. I was a city girl, a Jew interloping in a world that didn't belong to me, and thus never dreamed of questioning a family with such an equestrian pedigree. I didn't deserve to be in that world, some part of me was convinced, and Andre was happy to remind me.

On an evening when I was eighteen, I walked toward a dais through some sort of rinky-dink American Legion hall, packed with equestrians in prom dresses eating salmon terrine at the Long Island High Score Association year-end awards gala. My name was called, and I scraped back my chair to accept the division championship for the entirety of Long Island. Andre grabbed my arm and hauled me toward him, his mouth almost against my ear. "It just means you sucked less than everyone else," he said. For decades, I agreed. There's a photograph where I am holding a ceramic trophy and a drooping tricolor ribbon, taken a few seconds later. I still can't look at it.

Andre told me my only prize at the Hampton Classic would be some fried dough. He made real my worst fears about myself: that I was nothing. Mostly his cruelty took the form of an overarching roughness with animals and people.

I accepted the goings-on around me at the green-trimmed barn on Oakview Highway as methodology, believing for almost a decade that toughness equated with rigor. I was also eager to please Andre for myriad reasons I have only unpacked as an adult. In part, it was because he was handsome in an overly sunned, Robert Redford way; here was the All-American father I desired over my oddly named and accented, opera-obsessed dad, who instantly outed me as Other the second he showed up at a school function.

Striving to please cruel men has been a hallmark of mine. Through my near decade of blind commitment to Andre I have come to see that in me that impulse bordered on pathological.

It is a deeply inculcated way of being in the world that has been the project of my adulthood to unwind. Knotted into it are my no-brothers: men who are supposed to love me but don't are the first men I knew. Andre was merely a replication of those first not-loves, the people who loathed me by dint of me being me. There maybe even was a cold comfort in the familiarity of that feeling, that affirming in me of a baseline setting of un-worth.

Horses were so appealing because they loved me and stayed close to me no matter what, even as the three human men who were bound by blood to do so declined. It was and is a searing ache that was my normal, and so I sought out vicious men, whether as coaches or later as lovers, whom I could convince I was worth it, endlessly replicating that familiar pain. It was my reality, and for decades, some part of me thought I must have deserved it. I could never recall what crime I had committed as I pondered it as a little one, curled up behind that swinging kitchen door. But I was sure I had done something to warrant a trio of family who didn't want me. I was sure I deserved their cruelty. And Andre's.

On an afternoon when I was twenty-one, I was driving back to East End Stables in my mother's silver SUV when I suddenly turned the vehicle down Gerard Drive, a spit of land in East Hampton. There, marsh water eddies on one side; among the beach grass are platforms built atop telephone poles for osprey. The cresting bay bashes the rocks on the other side of the spit.

I gunned the engine and raced breakneck down the prom-ontory. I was so enraged by a wave of realization that my foot jammed hard on the pedal and I almost swerved off the penin-sula. It wasn't my fault that I had entered my brothers' story.

My birth was not an offense I had lobbed at them; neither was the divorce that fractured their family and gave way to my own.

Then another wave: I had spent eight years riding with Andre because he was my objective correlative for those three men, my brothers, whom I would never please, no matter how good I was to them, no matter how many championships I won. That day I released myself from all four men.

I never set foot at East End Stables again.

SHADER

It was so dark I couldn't tell where the gorse ended and the horizon began. In the Hither Hills of Montauk, the dark was like snow: a soft, dampening quiet that blanketed everything so completely it almost seemed like something you could touch.

It startled me, that deep dark, and I realized that this night was the true, impenetrable kind that I'd never found in New York City. There, the dark is perpetually neon-tinged, buzzing like a faulty halogen. In my city, that night light never ebbs; it's simply glowed-away by dawn. I was in high school at Brearley, where I had recently learned of the city's unwinking night first-hand as I tried on the cloak of a new identity: bad.

I'd spent nights that school year perfecting the art of tip-toeing past my sleeping parents and the Cerberus of doormen. Beyond them was a world of bleary sunrises squinted at from Lower East Side tenement rooftops. Weekday nights and week-ends, I'd pad through the marble lobby of our apartment build-ing barefoot so as not to let the flat-capped doormen hear me click over the stone. They were the watchful big brothers that my own were not: once, when I splattered myself out of a taxi, my favorite doorman, the one who had the whole building's birth-days memorized, carried me over his shoulder into my empty apartment and tucked me in, my coat and shoes still on. I gave him a doughnut a day for a month for not telling my parents.

At the door to the street, I would sling on stilettos my mother didn't know I owned and dive into a waiting yellow cab of girlfriends. They were fellow private school ne'er-do-wells and girls I'd met astride ponies in Long Island who had grown up too. The night was a new thing to us all.

The year I turned eighteen, when I was not out on Long Island at the barn with Willow, I was in the thrall of that most urban snake charmer: the party promoter. Hired by elite clubs, promoters are supposed to get out the word and fill the venue. But let's be straight here: they don't promote parties; they pimp them. Their true purpose is to pump events full of young, beautiful girls. They'd meet us at tamer parties, house parties thrown by us high schoolers, and scrawl our numbers in little black books. Our phones would ring later with their invitations, summoned for the sole purpose of accessorizing a nightclub. We didn't know enough then to mind.

I was paid for my time in glamour: entrée to the city's elite night world, no ID necessary (useful, because few of us were anywhere near twenty-one years old). To get in the places of the era—clubs like the palm-tree-lined Bungalow 8, where your camera was confiscated on entry lest you snap a celebrity, or Tunnel, where there were no rules at all—men then and now had to buy access. Tickets to the party came in the form of overpriced bottles of Cristal or other alcohol. If they put up five thousand dollars for a liter of the stuff, a man could swish into any club he liked.

Girls—and I use that word with intention; we were certainly not yet women—always got in free.

That Cristal would sit on a small table in a throbbing club corner, surrounded by bottles of vodka and fluted carafes of cranberry, orange, and grapefruit juice. And us. We were permitted to partake as much as we liked, which was always too much. Ensconced in pleather banquettes with our gratis booze, we were expected in return to spend the night sitting beside the old men and being seen.

Looking back, I realize what fire we played with and how deeply toxic the whole arrangement was. The promoters lured us underage girls into places we had no business being, to spend the night seated beside predatory men. Entranced by access into the nightlife of adults, I stepped excitedly into what I now know were honey traps designed to ensnare me into much darker things than a night of dancing. At the time, naively, I felt cool. Now, I'm furious.

One night at a club in Chelsea, I pushed the envelope of the arrangement. As the music jackhammered, I grabbed a girlfriend seated beside me making small talk with one of our hosts, took her by the hand, and threaded her away from our table to the dance floor. We wobbled and danced with abandon—for about three seconds. The party promoter who had invited us snaked his head between us. "Dancing?" he hissed into the pulsing nightclub with schoolmarmish horror that made us burst out laughing at the absurdity of the question.

He was irate. We were scuttled back to the table full of older men; our job, he reminded us, was to decorate it. He put his hand in front of my face the way you would before the wet nose of an overeager Labrador. "Now, stay." I sat miffed for a moment, then grabbed my friend and darted back to the strobes and bass lines of the dance. Again, the promoter steered us by our elbows back to our table. The next time we tried to leave, the promoter was a step ahead of us—and he had a bouncer with him. They hauled us back to the banquette, and the burly guard stood sentry. He pressed his fingertips into the flesh of my arm, holding it down to the pleather seat. Hard.

Before that night, I had been thrilled to find myself among the dazzling girls sipping too-strong drinks in the city's glam dark. I was secretly enthralled by the power my new teenage body wielded and by the fact that, at least outside school, I was a commodity. But suddenly, on the sticky faux leather seats of the club, any veneer of glamour was off. The pay-for-play nature of the glitzy arrangement had been laid too bare to be ignored. I was disgusted. And scared.

This time, when we again managed to duck away from the bouncer's gaze, my friend and I ran out into the street, forever. My club days were done.

In the summer, my passion for horses pulled me back from whatever precipice over which teenagers always seem determined to hurl themselves. You couldn't ride hungover, not if you wanted to win, as I'd found out during one horse show early on in my brief bad phase. In the dusty sand ring at a Long Island competition the summer following my discoveries of my cleavage and tequila sunrises, Willow blitzed through a jumper round as dogged and willing as ever. And I steered her in exactly the wrong direction over the pattern of jumps—not once, not twice, but in three separate rounds.

Andre walked away from the in-gate before I strode out. The sight of his back and the sun-charred nape of his neck was more distressing than his usual craggy scowl. He wouldn't look at me. When I dismounted back at the trailer, it was the first time he spoke.

"You ride like that again," he spat, as I blotted the sweat darkening Willow's white fur, "and I'll tell your mother exactly what you did last night."

I had of course not breathed a word of the beach party the night before. My pony pals and I had spent the night sucking back cans of Natty Ice to impress some local boys. They had driven us over miles of beach in a pickup truck one of the guys had borrowed from his fisherman dad to a party around a hidden bonfire. Shells crunched under the treads as we bounced toward the bonfire, glowing through the windshield miles down the shore. My girlfriend and I squeezed hands in the backseat, in disbelief that us city girls had wound up in a country song—and that our parents didn't yet have an inkling we'd grown up.

But I didn't have to tell Andre anything about my wild night; my riding did that. It was the last time I rode hungover.

Precluded from booze and disgusted with Manhattan's nightclubs, where I seemed only to be collateral for a bottle of Cristal, I needed a new way to be bad. That was how I found myself wandering through a moonless night over the hillocks of Hither Hills State Park, 1,755 acres of pitch pine and sun-crisped seagrass in Montauk. There, Long Island narrows to a tip in the ocean, into a spit of sand and scrub that locals simply call The End.

In my hand was a lasso. I was there to steal a horse.

* * *

Dark on dark, all around me in that blackness were the blacker silhouettes of resting horses. They were the sixty-five head belonging to Deep Hollow Ranch, a dude ranch owned by Rita and Rusty Leaver, where as a teenager I worked as a summertime cowgirl. It was more lucrative than grooming at East End, which had paid me only in riding lessons, but not enough to support my horse habit. So at night, I tended bar at a restaurant in town and forked over my earnings from both jobs each week to help my parents support Willow and pay for horseshoes and tack and trips to shows. I'd ride her in the morning and then head to my day job as a cowgirl in Montauk. From 8:00 a.m. to 5:00 p.m., my gig was to take terrified tourists on horseback rides lasting two and a half hours. They were long, sun-baked strolls over the salty land, where the wind whipped in through jagged pines from the Atlantic Ocean.

Deep Hollow has an illustrious, unexpected history for an equine outfitter in a posh beach town: it's the oldest ranch in the United States. In fact, it's older *than* the United States, with records that cattle were kept there on land leased from the Montaukett Indians as far back as 1658, and probably earlier. When I stop to think about it, it makes sense: cowboys and cattle may be tropes of the West, but they had to start somewhere—Out East. The plains that would become Deep Hollow were communal cattle grazing land then, overseen by "keepers," prominent local men who lived on the land as stewards and kept the peace and the pasture. The Montauk keeper was the first cowboy, you might say. In 1898, it was a military base for Teddy Roosevelt's Rough Riders, the scrappy cavalry stabling on the same fields where the sixty-five mustangs and others who made up Deep Hollow's herd got fat all summer long.

By the 1900s, Deep Hollow was already a western-themed resort, much the same as it was when I spent my dusty summer days there. Back then, I wouldn't have been able to set foot on the land. A 1930s brochure for the place, unearthed for me by a librarian at the Montauk Library, bears the ugly stamp: "The clientele is restricted to Gentiles."

Most of my customers were in abject terror for our entire slow-motion journey to the sea. Thing is, over a Bellini on the beach, an afternoon playing cowboy sounds like a peachy idea. But once we got swinging out over the scrub brush, the hack line of animals marching under relentless sun, any buzz was soon replaced with a sticky hangover.

As idyllic as it looked, by the time we got to the midpoint of the trail—where from between the beach plum bushes came the glittering view of Block Island Sound and the horses would leave the pine barrens and stride out onto the pillowy sand—the tourists would be in agony. Customers rarely wore riding gear. In swim trunks, their inner thighs blistered and pinched against the stiff saddle leather. An hour of that pinching, and most of my clients were ready to kill, and not just the fog of gnats that took every tour with us.

The real problem had nothing to do with chafing or bugs. The truth was that most of the customers never really seemed to like horses; they just loved the idea of them.

I think the riders might have been more terrified than irked had they known the truth about my sixty-five charges: the scrappy creatures were barely broke to ride. The hack line horses were in fact trained just enough to know that their job was to plod with their muzzle up their fellow soldier's rear end. Try to turn away from the herd to retrieve a tumbling cell phone?

No can do, my horses said. Halt while everyone else is marching off, to disentangle a rider's ponytail from the clutches of a spiky rosa rugosa bush? *Not on your life*, my horses said. (By the way, that rosa rugosa beach rose kept that blond lock in its stems for two summers, a flag fluttering in the sea air as a reminder of just how callous my herd could be.)

The horses were almost feral. How could they not be? The whole herd worked only three months a year, during the resort town's high season. They kept time with the well-heeled hordes that escape the thick New York City heat for the Hamptons hamlets each summer. The horses spent the rest of their lives fallow in the oceanside fields. As the snow clumped in the fir forest around them and they chewed for beach grass through the ice, they would forget a little more their roles as beasts of burden. By the time we pulled the animals from winter pasture for another summer season, they had other ideas about what they were supposed to be.

They were only halfway decent by Memorial Day because of Jason and Jeremy Hess, two brothers who were real cowboys imported from Indiana to manage the farm. Jeremy was soft spoken and scruffy. He had tobacco-colored teeth and eyes that glinted like the tellin shells we crushed under hoof on the beach. They seemed to flicker more whenever we got into daredevil contests over who could do a more acrobatic dismount. And they positively glittered when, after attempting to outdo him with a somersault off a horse's rump, I would inevitably end up in the dirt.

Jason was redheaded and perpetually reddened by the sun, with a fringe of red eyelashes around his light eyes. I know that because I studied each lash furtively whenever we scrubbed the watering troughs together, hoping he wouldn't notice as he bent over his work. That summer I thought a lot about his bright smile. And how when he spoke of horses he loved back home or fondled the ears of the ranch's free-roaming donkey, his red state toughness would evaporate entirely.

The Hess boys showed up early in the season to work with the herd. It was a pro forma regentling, a boot camp in which the brothers strapped saddles on the animals and declared them trained. They were still mustangs in May, and we always hoped for a June heat wave to simply subdue the horses with sun.

Despite its exclusionary past, Deep Hollow never said no to a prospective client for any reason in the modern era. When a rider who weighed more than three hundred pounds arrived every so often, a weight that can be a challenge for a horse to bear, Jeremy would whistle to Jason. Out would come the ranch's stretch limo: either Bill or Barney, a pair of stout Belgian drafts who usually pulled a hayride wagon over the hills.

Bill and Barney were also our saviors. The Hamptons are glamorous and gorgeous and glitzy. They are also hurricane prone, nestled on an exposed finger of island where the Atlantic blasts the eastern shore and bay waters lap its western coast. Every summer at least one hurricane would hit us, knocking out power and washing out the roads with seawater that slunk across the pavement.

After one big storm, the electricity cut out for days, and at the same time a heat wave boiled over. Our hoses were dry. With no power to the pump, we couldn't water the horses, and the gas station nearby couldn't pump gasoline to fill the trucks to go get water. To conserve fuel, we hitched up the Belgians to a hay wagon. I climbed into the driver's seat in Wranglers and a bikini top. Atop my head was my favorite straw cowboy hat that made the real cowboys Jeremy and Jason deride me as "Beach Barbie." I flapped the reins on Barney's and Bill's broad backsides. The team clopped to town on a mission to fill barrels at the firehouse on Flamingo Avenue and tow the casks back to their thirsty comrades at the ranch.

As we trotted back home down Montauk Highway, I passed John's Drive-In, a retro diner that whips up homemade ice cream. Suddenly a woman sweating outside flagged down my wagon. "We can't get the generator running for the freezers," the manager said in a panic. "Do horses eat ice cream?"

Into the hay wagon went four huge drums of melting ice cream. I swished the reins and Barney and Bill cantered back down Route 27 to the ranch before the treat turned to soup in the sun.

I never did tell her the truth: horses don't eat ice cream, but ranch hands sure do.

The days at Deep Hollow were hot and grueling; even just walking the undulating trails was tiring work for the herd. The horses went out on ride after ride, eight hours a day, earning their high season keep for the dead winter. Jeremy and Jason served as each excursion's taxi dispatchers, pulling on bridles, cinching up girths, and getting newbie riders aboard. They shepherded wobbly customers up an extra-tall mounting block that came up nearly to each horse's shoulder so that the elderly or the chubby could gently step on. Jeremy would look each rider up and down as he or she stood on the block, waiting to mount up. He'd assess girth and gut, height and heft to match each one with a suitable mount.

In reality, Jeremy's assessments seemed to have had almost nothing to do with a rider or a horse's physical size. With his rough handshake and quiet eye, what Jeremy was assessing, it seemed to me, was human and equine personality: any hint of bravado from a rider, and he'd bark out for Jason to bring up the oldest, lamest of the cohort. He knew that rider's machismo would inflame any of the less docile members of our troupe. He would also do the opposite. When an autistic girl with crooked pale pink spectacles came up one afternoon, Jeremy matched her to a horse who was a notorious tugger. The horse would practically pull arms out of sockets as he treated the ride like one giant drive-through, snacking on every passing leaf. The match was cowboy courtesy: by the end of the ride, the tiny girl had prevailed over the gluttonous beast. When she trotted down the trail back to the pack in the pen, victory glowed beneath her sunburn.

To each client, the Hesses disseminated a four-minute lecture on how to ride Western. Just as there are different types of transmissions, automatic and manual, there are several different modes in which to ride a horse. The main two are called Western and English. Cowboys are the ones who ride Western, in a wide saddle with a plank-like seat. It is a secure throne for long vigils in the tack pushing cows across the landscape. At the front of the Western saddle protrudes what is called a horn, an anchor for a hog-tied steer, say, at the end of a rawhide rope. But most distinct in Western riding is how you steer.

Western riders use the pressure of the rein against the animal's neck to guide it. No more than this whisper of leather against fur is needed to instruct: horsehide shimmers with nerve endings. Watch a mare at rest in a field, and you'll see her skin is so sensate that every few seconds, it trembles at so little as the six teensy footfalls of a horsefly.

In English riding, my preferred discipline, the saddle is slim and holds the rider close to the horse's body. The steering is different: the reins in each hand, it's a gentle pull left on the bit in the animal's mouth to go left, right to go right. In both, delicacy is paramount, but particularly in riding Western. The typical cowboy bit is a hulking, cantilevered contraption. If used incorrectly, like all bits, the metal can do cruelty to a horse's gums.

"Reins in one hand, kick to move forward. To stop, pull back," Jeremy and Jason would say, demonstrating with a bridle hung for the purpose on a fence post. Then in unison, they said what they always said to clients who could unwittingly use a bit so brutally they'd flip a horse over: "*LIGHTLY!*"

My guide horse was a chocolaty black gelding named Shader. Guides rode both the better horses—and the worse. As unbroke as our entire hack line was, the guide horses were typically even more rambunctious. For the most part, the lead horses were not bad horses at heart. It was just that stationed at the head of the line, storming through the woods, the leaders were freaked out. Most horses are in their comfort zone following the leader, programmed in their chromosomes to look to the lead stallion and head mare to keep them safe.

We never asked our chosen guide horses if he or she was an alpha, eager to lead the way, or if they were a beta, whose happy place was up another horse's butt. Those beta horses unlucky enough to be underneath me at the guide position would return from each trail steaming and sweating, even if it had been at a breezy walk. Whatever beta horse you put in the lead role would spend even lazy rides with muscles, ears, eyes, and nostrils vibrating, vigilant for danger. For the alphas, leading their band as they were born to do, their hides stayed cool. I could tell if I was on an alpha instantly: as we marched out from the corral, I would feel their chests beneath me swell with something like pride.

Shader was an alpha, and his past had made him brave, eager to dart ahead of the pack. He was a breed known as a Standardbred, just like the pair who years later I would watch trot across the ballroom of the Waldorf Astoria. It's a brawny, thick-shouldered horse with an erect head carriage developed in the 1700s for the sport of harness racing, pulling a flimsy carriage around a track at speed. He had done the rounds at a track in New Jersey in his younger years, but all his boldness couldn't make up for his racehorse failings. He was just plain slow. Failed track horses often end up in Canadian or Mexican slaughterhouses. Shader had good feet and a good mind; he was one of the lucky ones who found himself a second career. Despite his lethargy in races, generations of breeding made Shader want to pull out front of any other horse in his radius, like a good racehorse should—so much so that whenever he was stuck behind another in the hack line, he was known to trample, seeking a finish line that was in his head alone. Shader may have been a born loser, but he was also born to guide.

For most of our journeys, the horses planted their noses in between each other's cheeks. That way, every time the horse in front swished its tail, it would also swat away the gnats sipping at the corners of the following animal's eyes. Out in the wild or in a backyard paddock, pairs of horses often stand side by side, nose to tail, mutually fly whisking.

Halfway down the trail to the beach was a rise and a clearing. I would stand Shader atop a bluff, and as one, the hack horses would take a pee break. Across the spit was an old military base, where a World War II–era satellite the size of a Mack truck rusts overlooking the ocean. At the overlook, I'd roll out my patented, hammy guide joke: "Coming up is a part of the trail called 'Oh Wow,'" I'd shout back over my shoulder to the riders fanning out behind me down the path. They'd be puzzled, until their hooves hit the jaw-dropping rise and caught sight of the giant rotting radar dish over the bluff, and a chorus of "Oh, wow" would echo down the hack line.

It's called Camp Hero, a former military installation, but the tower is long abandoned, the acreage turned to parkland. The monstrous building is a warren of caverns once packed with radar equipment. Around it, gunners once trained cannon west, ready to hold at bay a naval attack by the Nazis. The structure malingers now, shrouded in rumors that Camp Hero was the Hamptons' own Area 51, the building full of operating tables on which aliens were vivisected. There aren't any operating tables inside. I know, because I checked personally when I broke in with Jeremy during those bad girl days and we looked around. It was grand, graffitied, and imposing, but up on the roof, the radar dish no longer looked otherworldly as it had from the saddle.

That didn't surprise me; everything looks better from a horse.

*　　*　　*

From three summers of observation from the back of my lead horse, I can say without reservation that men were the worst seaside cowboys. Not all were, of course, but on days when one of my trail rides went south, when there were spills in the sand or I had to thread Shader through the underbrush to retrieve a bikini top a bayberry bush had snagged off: *cherchez l'homme*, look for the man.

Aping silver screen cowboys, men pulled hard on the reins in an attempt to swashbuckle, cranking the heavy artillery of the Western bit in the hack horse's mouth. They kicked Teva-clad heels into my horses roughly. When an animal became unruly, men most frequently went at the creature harder rather than softly deescalating or turning to me for help. When a horse took over, hustling after its pack-mate, men rarely went along for the ride. Instead, they hauled back far too hard to stop him, heedless of the metal on the horse's gums. More than one guest nearly flipped the entire horse over this way as the animal sought to avoid the pain in his mouth by rearing skyward.

Worst of all was when men showed off. Eager to make it look easy, they frequently would grab their girlfriend's reins to haul her horse along at a faster pace. But the horses, threatened by being crowded on the narrow path, would aim teeth and hooves at each other. Just as often, men would drop the reins and stand in the stirrups to strike a pose for a photo, only to plop off the side when the horse used the opportunity to head into the bushes for a snack.

Steeped in workaday machismo, it seemed to me that these men often tried to use the same world-beater tactics of their Manhattan boardrooms on the horses. One problem: my lightly trained animals had never received the memo they could be ruled.

* * *

With Camp Hero illicitly conquered, the Hess brothers and I had new mischief to think up after the last client left for the day. It was dusk in the barn as we curried the horses and stowed the saddles in the hayloft when they came to me with their plan.

That night, we were going for a joyride. We were going to steal horses and thunder through the night across Montauk, all while the ranch owners, Rusty and Diane, slept in their house across the street. Key to our success would be the continued somnolence of the Leavers' pack of hypervigilant border collies. The dogs knew at least twenty different commands, all spelled out in different pitches of Rusty's looping whistle, and they lived and breathed to protect livestock from thieves like us. Would-be horse rustlers, we couldn't rustle a leaf.

The ranch staff met back at the stables at midnight at the Hesses' instruction, in silence, lights off, and headed into the pasture. It was filled with night noise, cacophonous cicadas clinging to the silvery-leaved Russian olives and puffs of shadberry. The collies couldn't hear a thing over it. I stalked into the dark field behind the barn, with a coiled length of rope, to where the horses dozed. I knew the hills by heart that August night, something I was surprised to realize. I was a city girl stepping through blackness. But the invisible acres pouring out before me were as familiar as my bedroom in Manhattan, as if I was merely tiptoeing across at night to fetch a glass of water.

Somehow, we each found horses in the night, led them back to the barn, and saddled the sleepy animals in ninja-like silence.

Then we stole them.

Horses have some of the largest eyes of any land mammal,

168

outsized eyes adapted for the dark. They have more rods than cones, and a structure called a tapetum lucidum—that thing that makes deer eyes and cat eyes iridescent in headlights. It makes their scotopic, or low-light, vision superior to ours. Horses see in color, but not as robustly as we do, as any rider who has sat on a horse spooking at a garden hose that the animal is convinced is a cobra can tell you. But in the dark, compared to us, they see like owls.

That night we rode faster than I'd ever before ridden in my life in broad daylight. We were confident of our horses' fortitude and their sharp, self-preserving wits in the black. The herd's almost-wildness that plagued our workdays and sent tourists home with sand in their britches regularly was jet fuel. The ten horses who shot through the night forest knew the rocks, the trees, and the trail. The animals had spent winters as horses in the truest sense, beholden to no one but themselves. In drumming rain or sifting snow, the crannies of Hither Hills had become their kingdom.

That night they were royal.

Their tattered tails were no longer fly whisks but heraldic banners, streaming their glory behind them. They galloped on one another's heels, crowding down the paths, each horse goading the line on more sharply than any wicked spur. In daylight, these animals plodded; that night they whipped through the roller coaster of tracks as deft as dancers, as surefooted as sherpas. They slalomed between the scrub oaks as lithesome as eels disappearing among coral.

"*LIGHTLY*" was Jeremy and Jason's command for how to control these horses during the day, but in the dark, the horses controlled the ride. I dropped the reins and clung to the pommel horn as we clattered down a stony slope. As we pounded down the miles to the beach, I found myself uttering a barbaric yawp with every dip of the terrain. Far from the dogs, I yelled out in praise of something larger and more powerful than myself, that the wild, endless strength beneath me let me access: freedom.

When we got to the bluff of Oh Wow, the horses automatically pulled up. They were used to stopping to let the tourists pause for pictures there. But that night I imagined the horses, unencumbered by rules that demanded their supplication, saw for the first time the startling view. We stood just like I fantasized the gunner atop Camp Hero once did, scanning the sea, poring over its details in a night without a moon. The Hesses had chosen the cloudy night for its cover of darkness, but at that moment, it seemed to me that moonlessness let me write my own story out into the blank canvas of the dark—the way a birthday child squeezes her eyes shut when the candle is blown out to truly envision a wish.

I was higher than any nightclub could have made me, more electrified than on any dance floor. The baubles of the urban night that had enticed and wounded me dimmed with every hoofbeat. Sure it was stupid—a single misplaced hoof could have sent a horse tumbling and the entire string cascading after—but so was doing shots with lecherous men and spinning out into a city that sometimes sucked people down whole. None of my Manhattan rebellions had ever made me feel so free.

On the beach, I dismounted. The other staff scattered. Some raced each other on horseback in short bursts. Some came together to hold each other furtively in the dark. I cast about for Jason, the Hess I wanted to kiss.

A few days before our midnight ride, I had worked up a summer's worth of courage to kiss him square on his plain red mouth. We were in the far-back field behind Deep Hollow, in the battered old pickup truck full of hay, dumping lunch out for the horses. Jeremy stood in the flatbed of the truck as we drove slowly over the hills, pitchforking out the hay to the salivating horses every few feet. I sat in the cab as Jason jounced the truck over the hillside, striking poses under my flimsy straw cowboy hat. I batted every eyelash I had as we cruised around the field. The herd studied us with hungry eyes. Jason studiously declined to take notice of me.

Then I grabbed him by his plaid collar and kissed him deeply. The road-weary pickup creaked to a stop. In the bed of the truck, Jeremy half-tumbled out with his pitchfork as the jalopy jerked to a standstill. He scrambled back and peered through the back window into the cab at our clinch. With a tremendous clang, he drummed his pitchfork on the roof of the sorry vehicle. I nearly bit Jason's lip with the fright, and we spat each other out, laughing.

But Jason had more romantic interest in the ranch donkey or a pitch pine—or in anything—than he did in me. Today he is married to the Leavers' daughter Retta, actually living out a cowboy and cowgirl romance. So on the beach during our illicit nighttime ride a few days later, Jason was suddenly nowhere to be seen; in fact, he seemed to be taking cover from me in the night.

Briefly crestfallen, I picked out the fir needles tangled into my hair from the mad dash through the trees and led my horse to the edge of the water, sitting down in the sand. Around me I could barely make out the silhouettes of cowgirls and cowboys scattered like gulls down the beach line. Spent from the electric gallop, my horse stood quietly over me and touched my hair with his lips, sampling a strand in case it maybe had turned into hay. Finding it hadn't, he breathed deeply and looked out with me at the wine-dark sea.

I remember that moment more powerfully than any kiss I've ever had.

SAMSON

I strode across the auditorium at Brearley with a clink and a jangle, my spurs jingling across the polished floorboards, my wooden nightstick thwacking my hip in rhythm from where it hung from the holster at my waist.

By senior year, I had long given up trying to reach the level of searing exactitude demanded by the elite school. It was a pressure that packed the bathrooms between class with girls purging their focaccia pizzas and anxieties. The women I met there have some of the most incredible hearts and minds I've ever encountered, and they remain so to date: they run governments and banks, theaters and newspapers. Our experience of our school differs: some fit, some were misfits; we all felt the pressure, yet only some of us cracked.

I did.

At Brearley I failed: for being late, for handing in crumpled papers, for speaking out of turn, for answering questions I thought were more compelling than the ones on the test—for things that I know now, as a professional writer who still can't spell, have nothing to do with apprehending information and everything to do with my inability to fall in line. My competitive high school was my weekday version of my weekends spent riding with Andre: a constant reminder that I wasn't enough, that something about me never measured up. Some subconscious part of me was (is?) addicted to that feeling.

My grades by the end were abysmal. I had been sidelined to a math class so remedial that even the teacher went along with our nickname for it—id'jit math—where the grade's six or so designated misfits sat humiliated as we were taught something like fractions, again. That year, I received close to a D in English and yet a week later earned a perfect score on my English SAT. Instead of feeling like vindication—"I know I can't spell, but see here, the College Board says I'm smart!"—it made me realize that in school, what I was truly being graded on was being a Brearley girl. I had failed. I couldn't make myself be Mehitabel, that fictional ideal girl, and Sarah didn't pass muster.

But that afternoon in the auditorium, I was at last someone else. I was there to give a presentation on how I'd spent my final year of high school. At the front of the room, I unholstered my nightstick.

"Officer Nir," I said before the assembly. I was dressed in my black riding boots and officer's britches, green with a yellow stripe from waist to ankle, all topped with a crisp taupe button-down embroidered with the crest of the seal of the City of New York. I turned to my assembled classmates and teachers: "New York City Parks Enforcement Auxiliary Mounted Patrol, reporting for duty."

* * *

Chris was the first mounted unit horse I ever met. He was the color of a new penny and seemed to have a city slicker swagger.

He was retired from the force in 1998 and sent to an exurban life at East End Stables in East Hampton, where I rode. We'd joke at the barn that like many Manhattanites, Chris too summered in the Hamptons. But Chris was done with that gritty city life and never going back. With his workmanlike build, he looked profoundly out of place among the show hunters in the Hamptons. But he was trained to tolerate anything from the burst of flare guns to the jostle of crowds, so he could certainly take the occasional jeer from his human stablemates, some of whom hissed that an urban quarter horse did not belong there.

There's an old saying used by horse people to which I ascribe (and that I'd like to see more of in the wider world): "Pretty is as pretty does." To perform beautifully is to be beautiful. There are some people who dote on horses for a particular color, for a splash of chrome, the sweep of tail, or a patterned hide. I have been known to salivate over Knabstruppers, a polka-dotted Danish horse out of Pippi Longstocking's daydreams. But then I remember the mantra: pretty is as pretty does. Admirers of equines merely for their beauty or a particular flare are generally seen to be missing the point. Horses are workmen, athletes, adventurers, accomplices. How they look is inessential. But how they go, how they are—in other words, *who* they are—is what makes any horse beautiful.

Chris, with years of service in the New York City Parks Department as a ranger, was a stout, handsome horse. But by this other measure, by pretty is as pretty does, he was drop-dead gorgeous.

He was owned by Sara Hobel, a magazine executive who

had come across Chris's colleagues when she was in the midst of a midlife crisis. More precisely, she was in the middle of a jog in Central Park. Running past park rangers patrolling on horseback in 1996, she commented aloud: "That looks like a really fun job," the first spark of what was to become an existential equestrian revelation. The ranger who overheard her remark invited her to volunteer with the force, and Hobel was hooked. She quit her job, became a trainer for the mounted unit, and eventually the director for the rangers citywide. Chris was her mount, a bold member of the park force for four years, leading drill formations and patrols. But like Amigo, he had a single debilitating weakness: his was flags. "He would grab the bit, throw up his head, and bolt at a gallop back to the barn, which usually was across major streets," Hobel wrote in Chris's obituary, when he died at age twenty-five. "And needless to say, he did not wait for the light." To spare him the anxiety, she adopted him and retired him early Out East, far from fearsome flags.

"I made a quarter of what I did in the corporate world," Hobel would say to me and others often, reminiscing about her career when we stroked Chris's copper nose together at East End, congratulating him on a job well done—as best he could. "Happily."

At Brearley, second-semester seniors were permitted to take on a job, for which we would receive school credit. The rare laxity of the Brearley regimen seemed born more of necessity than because of some sense that joining the workaday world could be edifying. That's because once college applications were in, it was all the school could do to herd my burned-out posse of miscreants away from nearby Carl Schurz Park and back to class. The leafy park was a few blocks uptown and at the foot of Gracie Mansion, the buttercup-hued residence of the mayor of New York City. Carl Schurz was where we told teachers we spent our afternoons convening meetings of the PAC—the Park Appreciation Club. PAC consisted solely of burning through entire packs of American Spirit cigarettes in a sitting.

One afternoon, my throat chapped from too much Park Appreciation, I thought of Chris, the copper-colored ranger, and realized there were better ways to while away the day in the park. If I was allowed a job senior year, what about working for the Parks Department Mounted Unit? Hobel had often spoken about getting her start with the unit as a volunteer. Could I? One thing I did not think through clearly: if successful in my quest to join the mounted unit, I'd spend the rest of the year in Central Park chasing truants just like myself all the way back to school.

A few days later, my first experience with the mounted unit began. It started out a lot like the rest of my life back then: with me failing a test.

King and Samson were Belgian draft horses and weighed over a ton each. They were stabled at Birchbark's old home, Claremont Academy on the Upper West Side. At first it seemed unreasonable to me that the parks department would husband such gigantic creatures in the hive that was Claremont's upper story barn. But the icy blood that runs in the draft breed's veins helped them tolerate their lifestyle. City horses have almost zero time at liberty—what's called "turnout," those vital spans spent outdoors, cropping grass, riderless and naked, not working, just being. With real estate prices in Manhattan sky-high, it's hard to find a human a studio apartment, let alone the square footage for a paddock.

The truth is that life for the majority of equines in the city, from the hansom cab hacks to the patrolling police horses, consists of all work, no play, followed by a supper of timothy hay in a tiny box stall. Repeat. For breeds with temperaments that run hotter than that of plodding draft horses, a life without recess can make them understandably explosive. It's hard on the cooler-headed breeds too, but they seemed constitutionally equipped to at least bear it longer, like a cactus you can parch but only for so long. (But should you?)

I was back at Claremont to audition to be an auxiliary officer, an unpaid volunteer, permitted on patrol only in tandem with a real ranger. I would be a warm, horse-competent body to expand the slim ranks of the park's actual force of sixteen. The morning of my test, I was at last allowed up Claremont's foreboding ramp into its second-story barn to get my mount ready.

Up there it was otherworldly. Through a miasma of wood-chip dust slipped bars of slanting light, illuminating millions of particles eddying in and out of the horses' quiet breath. The light glinted through stall bars and spilled over the chewed-on Dutch doors. Some horses hung their long necks out over the Dutch doors into the aisle, creating narrow corridors lined with softly snuffling animals.

Shoulder to shoulder in the far corner at the back of the space, in a pocket of the barn filled with flecks bouncing in and out of the light, loomed a pair of perfectly matched Belgian draft horses. They were the park rangers King and Samson, and each animal was vaster than the other. Both were the golden hue of a frothy Lambic beer.

I dragged Samson out of his stall and got him ready in the narrow upstairs aisle. To reach the heights of his great back, I pulled a stepladder up to his round belly and wobbled up it with his saddle in my arms. Obligingly, he bent a head the size of my torso toward the ground so I could slip the crown piece of the bridle over his ears (or perhaps he was just rooting for the crinkle-wrapped peppermints in my pocket). Then dish-sized foot after dish-sized foot, he trod his way gently toward that infamous ramp. I led him to the precipice and released him to navigate the slope solo. The leviathan eased down the incline with surprising grace, placidly descending to where a park ranger in her full kit was waiting to catch him at the base.

I followed down the ramp a few feet behind him, eye-popped at the width of his blond rump and hoping the wooden slope would hold.

I carried the stepladder that I had just used down with me. I had to: the regular mounting block did not touch Samson's altitude. As I clambered up it and on, I felt my hips pop as my legs spread wide around his back.

Then nothing happened.

Sampson would not budge. Squeeze, kick, cluck, chirp, beg, cajole. I straddled Samson in the middle of Claremont's arena, urging him to walk forward with every iota of my soul and every ounce of my physical strength. The sergeant crossed her arms and looked up.

"The test has begun," she said.

* * *

Belgians are thick hunks of horses. The ideal Belgian has a flaxen mane and tail, a Fabio fringe edging a coat that ranges from butter-colored to rust. They are a massive breed, chill as stoners, and broad across.

The two equine rangers were distinct from each other in one essential way: Samson seemed as wide as King was tall. I remember King as 19 hands high. (The current record holder for tallest horse as of this writing, according to the *Guinness Book of World Records*, is also a Belgian. His name is Big Jake, a gelding from Wisconsin. He is almost 20.3 hands high, just a smidge over six inches taller than what I recall King was. King was not huge; he was gargantuan.)

As Belgians, King and Samson hailed from an equine family known as draft horses, a class of bulky, imposing workhorses whose most famous exemplar is the Clydesdale breed. You know them as the shaggy, bell bottomed horses kicking footballs or putting on displays of patriotism during the Super Bowl in Budweiser ads. Belgians are a particularly ancient race of draft, distilled from the broad-bodied warhorses of the medieval era. Drafts are a strain of horse distinct in personality and heft from the lighter-boned breeds almost the same way a somber bulldog differs from a restive setter.

Massive and muscled, their bones dense, and their feet Frisbee-sized. Of a type known as cold-bloods for their cool-mindedness, they are preternaturally chill. That demeanor makes them amiable and dutiful professionals at their job, and for millennia, that task was plowing. But tractors, and the burst of mechanization of the post–World War I era, sent draft horse populations spiraling down in America. By the mid-1950s, there were only two hundred breed registrations of Belgians a year in the United States. But never fear! Unemployment did not manage to wipe them out: today over two thousand are registered every year according to the Belgian Draft Horse Corporation of America, the country's breed group. Today they're useless but also somehow, it seems, needed.

There in the ring, Samson, true to his breed, was determined to chill. As our struggle grew more epic—me sweating, him not budging—a crowd of schoolgirls on a field trip to Claremont gathered by the dusty panes of the viewing room that overlooked the arena and gawked. Paul Novograd put down the radio handset and tore his gaze away from his call sheet of horses to order up, and stared, eyebrows arching. I hoped he didn't remember me from my Birchbark days, nor the fact that before this moment atop this behemoth, I could actually ride. Inside the ring, I felt the pressure. But Samson, it seemed, did not. I had been riding him for ten minutes now. Or rather, I had spent the past ten minutes sitting on him as he just stood his ground.

"Trot," said the sergeant. I lifted my legs on either side of him, raising each a foot distance off his massive flanks and let them fall on him with a thump. To say his ear twitched in recognition of the G-force of my heels hitting his sides would have been an overstatement. Samson remained utterly immobile.

"Try a walk," said the sergeant. Hanging from her belt loops was a wooden nightstick, and as much as I adore horses, the mounting humiliation made a small, wicked part of me wish I could turn it on Samson's fat bottom. He was unmoving, phlegmatic to the point of torpor, stock-still in the middle of the ring.

Instead, I did what I always do when horses won't horse: I pet the stupid blond crest of his stupid neck. He was a horse after all, and if he wouldn't comply and had just embarrassed me and ruined my future career in law enforcement, he still had being a horse to redeem him. "Good boy," I said (perhaps silently adding, *moron*), as I summarily gave up.

The ranger scribbled something in her notes and barked a final order, the only command I could actually obey: "Dismount!" I swung my leg over and stepped off Samson, right back onto the little ladder I had propped beside him. We hadn't budged. I was dejected. I had been riding since I was two years old and had astoundingly that day failed to achieve the very fundament of horseback riding: move.

Once I was off his back, Samson was instantly obliging. I led him toward the sergeant, and he strode at the end of the reins with ease. I handed her the reins, humiliated at a failure so rudimentary.

"Congratulations," she said, stopping my hot tears of embarrassment just as they threatened to fall. "You passed. Samson is a patrol horse, he knows how to patrol—outside—that's about it. In the riding ring, with most people, he just flies backward. You kept a one-ton horse from having its way with you, Sarah. Nice work." I reached into my pocket to feed that undeserving lout the first of an endless supply of peppermints.

"You can really ride," she said, extending her hand to me. "Welcome to the Mounted Auxiliary."

The Belgians were treated like giant babies, and as Samson's performance (or lack of) during my trial showed, they were accordingly allowed to get away with the antics of infants. The coddled mounted unit horses had a feed locker filled with so many nutritional additions that it looked like a medicine cabinet in a retirement home in Boca, packed with every quack longevity and vitality elixir in the market. There in the attic of Claremont, the unit had made a Canyon Ranch spa for King and Samson. It was totally unnecessary: the pair were healthy as oxen and could have lived well on air. But I think the over-the-top treatment was out of guilt for confining the massive creatures to the cage of the city. I felt it, and accordingly shoved an entire packet of those peppermints down the pair of them every single day.

At the assembly at school, my classmates would hear only of my glory, nothing of my mucking. "I walked a horse through fire," I told the rapt crowd in the auditorium, regaling them with the tale of a desensitization training we had done in the Bronx. The horses were taught to tolerate flare guns in case there was a riot and to walk over or through anything in their path. In Van Cortlandt Park in Riverdale we practiced prancing on plastic tarps and through curtains of streamers like you see in a carwash that would have sent Amigo into apoplexy.

And I told my classmates of the one and only time I put my nightstick to use. Sort of.

With a groan of exertion, a few days after I joined the unit that spring, I swung my leg over massive Samson and checked my gear. I tightened my holster a few notches around my waist with a tug. The thick belt of coarse black leather was slung low on my hips. On either side, it was hung with two structured pouches that thwacked against my flanks when Samson strode off. Where one might expect a pistol was instead a hulking black walkie-talkie and a leather-bound pad that officers used to write citations. (Because I was an auxiliary, the ledger in my holster was just a regular old notebook—but wrongdoers didn't know that.) Samson was clad in a kelly green saddle pad trimmed in yellow, matching the green of my britches and my stitched-on parks shield. From one side of the saddle hung a coiled green lead rope, from the other, a menacing wooden billy club.

All semester long, I strutted on patrol, cutting a formidable figure on my mammoth horse. As we passed by the Great Lawn, picnickers would skitter off to hide the rosé they had illegally uncorked outdoors, in violation of city law. Dog walkers, our main scourge, the ones who flouted the rules and let their beasts frolic unleashed, would scurry away at our hoofbeats, fearing the wrath of the officers' booklets of tickets. Reprobates getting frisky under leafy cover in the Ramble, graffiti artists with spray paint canisters poking out from their waistbands, all skedaddled before us. Truant teenagers scampered at the clop of my hooves and the sight of my baton.

Good thing they didn't know the weapon was useless. The inadvisability of sending a barely broke horse out topped with a barely trained teenager into New York City wielding a nightstick was not lost on the mounted unit. My knight's sword was forbidden to me, tightly affixed to my saddle pad and unremovable. It was utterly unusable and entirely for fearsome show. But an even bigger secret was the fact that Samson and King were disinclined to obedience; their intermittent steering, brakes, and reluctance to listen to the person hammering away on their back made targeting any lawbreakers and successfully apprehending them a profound challenge. Besides, were I to dismount these mountainous horses to mete out any law, how the hell would I ever get back on?

Cut an imposing figure? Check. That Samson and I could do. Almost anything else was beyond us.

My equipment all in place, I nudged Samson toward the ramp to the road, and he obligingly followed his towering partner King out of Claremont and onto Eighty-Ninth Street. Time for our first patrol.

As we walked, every so often my colleague, one of the actual rangers, would haul up and make the vertiginous drop from King's back to the bridle path. We were rule enforcers by trade, and thus condemned to be rule followers: every time one of our horses let loose a steaming pile of all that taxpayer-subsidized equine superfood, the ranger jumped down. She then unstrapped a dustpan from her saddlebags and scoop the heap up, depositing it among the hotdog buns and soda bottles in a municipal trash can.

I was lucky that as an auxiliary, it seemed too risky for a civilian to engage in the trapeze act of the poop-scoop dance. I watched as the ranger clambered down and ponied King with a green lead rope clipped to his bit, the other end tucked into her holster. He trudged in a circle after her as she bent over his pile, then shuffled along obediently at her side as she searched for a trash can. Then back again (horses this size took two scoops). Then she would search for a bleacher or bench to use to jack herself back onto King's back. I cringed whenever I realized Samson was also answering nature's call and the ranger had to dismount and do the dance on my behalf, all while ignoring the jeers of Rollerbladers zipping by our predicament on Park Drive. If only we'd left just one manure mogul on their path!

That first morning, we meandered into the park and, seeing no scurrilous people recklessly unleashing their bichons, strode instead down a pathway to a small patch of parkland wrapped in wire fence: the Belgian Boys' paradise. Unlike the other beasts crowded into Claremont, the mounted unit horses had a scrap of grass where we could let the team loose for brief periods of the day to graze and stretch. But I didn't unsaddle Samson as I led him into the pen in the park. He was far too tall to ever get tacked up again and just naughty enough to work body and soul to prevent it, so he and his buddy cropped grass in their gear. On the other side of the green wire fence, the ranger and I settled into the grass and stretched out.

Then: "Do you smell that?" the officer asked me, propping up on one elbow in the grass. I did. The tang of a very different sort of grass slithered through the afternoon air. It was 2001, and marijuana was not yet infused into gummy bears and lollipops or seen as a panacea for whatever ails you. Weed was bad. And who was bad in the middle of the day in the middle of the park? A group I knew all too well: truants.

"Mount up!" she barked, and we tugged open the gate to the paddock. I pulled Samson's bridle off the fence post and threw the reins over his neck too fast for him to dodge away, and before he knew it, the bit was in his mouth, knocking clumps of half-masticated forage from between his teeth. I dragged him to the edge of the fence line and, seeing the ranger somehow, miraculously (levitationally?) already astride King, I hooked my boot into the wire fence and used it as a wobbly ladder to climb aboard.

Without a word, we galloped through the paddock gate. We were in hot pursuit of the forbidden smell. Samson typically did whatever King did, and King was flying. I'll never forget the sound of the saucer-sized hooves as the ground beneath us changed from grass to the hardscrabble of the bridle path, to the tremendous clang of steel horseshoe on cement. We hung a right and a left, and their hooves clattered over the bluestone pavers around Cherry Hill Fountain. The roundabout was designed in the Victorian era as a scenic turnaround for horse carriages and the fountain as a watering trough for the animals. We pulled up and sniffed the air.

There, on the far edge, through the trees, I saw our quarry. Samson raised his head and flared his nostrils, thrilling with the unaccustomed bursts of movement slicing through his stalled and stabled life. A flicker of teenager darted among the elm. With a cluck and a spur, the Belgians put that Victorian turnaround to good use, whipping around the fountain after the teens.

I was stunned that Samson, all beef and irascibility, could move like this. Under me, he thudded and stretched, his toes gripping brick and powering his bulk across the park.

Deep in his bones and through his veins coursed the same thought processes that have propelled horses to fly through history: *It's safest to run. Being safe feels best. Ergo, running feels good. Really fast? Really good.*

Now imagine something predicated on that simple equation is packed into the upper story of a city building, living a life so constrained he even stays saddled in the field. Every pore of this horse was screaming to gallop. Finally, we were. As we careened to the far side of the plaza, Samson let loose the peal of a whinny. I think I did too.

Past the fountain, we caught up with the degenerates mis-spending their youth and cutting class—just as I did on days I wasn't ambling through the park on a horse telling others to do as I said, not as I did. The ranger touched King's reins, and he shambled to a stop at the edge of the plaza. Samson followed his lead. The kids were in a panic now; they were farther in the bush, and from what I could make out in the shadow and green light, it seemed they were receding more. Hand over foot, they began scaling the outcroppings of Manhattan schist that jut through the park, clambering into the heights to escape. They looked back at me over their shoulders. I touched the faux nightstick, unmovable at my side.

"Climb all the way back to school!" the ranger yelled, dropping her reins and cupping her hands around her mouth to make sure her admonition echoed up the crag. She laughed quietly to herself. She cupped her hands again. "Just say no to drugs!"

"You're not going to arrest them?" I asked her, incredulous that we had stopped our mad dash at the far edge of the Cherry Fountain esplanade. "We could go up that path and capture them at the top of the hill! We could come from different directions! Wait and set a trap! They have to come out of the bushes sometime!" She didn't respond. The ranger straightened her helmet on her hair and tucked a wisp back into place.

"Don't you think being scared shitless by galloping horses will get the message across?" she said at last. "Arrest them? And then truss them up and throw them across Samson's back? You want to try to convince him to let that happen?" I fingered his blond mane. She had a point.

"Plus, would *you* like that?" She looked me squarely in the eyes, and I cowered. My teenage brain instantly feared her. I spiraled, thinking rapid thoughts that quickly diminished into lunacy. Like: Was there a statute of limitations on truancy, or could I be arrested right there for cutting class the week before? And also: Did the ranger know? I started to speak. Stopped. Started again. Considered making a run for it on Sampson, now that I knew he was capable of speed. Then the ranger winked. "*If*, I mean, that were you?"

SWAMPER

By the time I was in my twenties, I figured that I'd discovered every last horse in New York City.

There are more than you might expect, and I'd fed a carrot to each, it seemed: from the hansom cab horses hoof-cocked on Fifty-Ninth Street in Manhattan to the trail mounts in Prospect Park, Brooklyn, to the hack horses who plod on the shore of Jamaica Bay in Queens. I'd visited the dressage horses pirouetting at a riding school in the Bronx, and even a donkey at a Staten Island Christmas pageant. I was sure I'd pet every equid muzzle in the five boroughs.

But it turns out I had neglected to look for horses in the middle of the East River.

In 2004, I had taken a year off college to work for a presidential campaign, and the candidate had spectacularly failed. I found myself suddenly demoted from a White House–bound whistle-stop bus tour across the West to my childhood bedroom. Grandma Frieda had died that year of pneumonia, but really from time. I had sat up with her watching *Jeopardy!* and eating coffee ice cream until the evening she passed. The next

191

day I decided to add her last name, Maslin, to my byline to have her near me forever.

One morning that year, I was reeling with a mix of crushed idealism, existential dread, and New York City's August heat. The answer to my malaise (as always) was to find the nearest horse, but I settled for a head-clearing bike ride. So there I was, pretending I was on a long gallop as I pedaled along the highway beside the East River that runs alongside Manhattan Island.

Maybe it was just the quarter-life crisis talking, but up by Harlem, in the middle of the water, I swore I saw a bright red country barn.

The conjoined islands between Manhattan and Queens, Randall's and Wards Islands, are home to a wastewater treatment plant and a mental institution. The entire island is enrobed in highway; Uptown, the East River becomes the Harlem River. On one side of the Harlem rushes the Franklin Delano Roosevelt Drive. On its other bank, the mighty Interstate-278 rises above on elevated stanchions. Did I really see horses?

I pedaled over the mint-green drawbridge connecting the FDR Drive and the island. As soon as my wheels touched earth again, I flung my bike under a laurel bush, vaguely hoping the shrub would conceal it, but too excited to care if it got stolen. I took off at a dead sprint toward the barn at the water's edge, across the island that echoed with the boom and growl of highway from every direction.

The stalls were shut tight, all but one: at the center, one Dutch door was ajar, its neat white trim bright in the hot sun. Every so often, the sound of exertion puffed from inside, followed by a shower of sawdust ejected out the open door. Someone was mucking the stall, and where there is muck, there are horses. My heart quickened. This was real?

I scurried up to the open door and peered around the jamb. "Horses?" was all I managed to say in my excitement. Perhaps it was more of a yell, because the person inside dropped her pitchfork with a startled yelp. "Sorry!" I whispered now as the woman bent and sifted through the shavings for her pitchfork, straightening up slowly with her hand pressed to the small of her back.

"I'm an old woman!" she said. "Scare me again like that, and you might be left taking care of all these horses!" She wore a bandanna around silvery hair that set off her café au lait skin and was pulled back neatly into a bun at the nape of her neck. She unwound the bandanna from her head to blot her brow. I still didn't see any horses, just a half-picked stall.

We stepped from the half-light of the stall and into the sun. "So you're looking for horses? In the middle of New York City?" she continued, impishly. I nodded, holding my tongue, still embarrassed at startling her. She began to walk along the side of the red barn, pausing at each shut stall to undo the bolt and swing open the half-door. Like jack-in-the-boxes, out of each popped a shaggy head, squinting in the morning light, hay breakfasts dangling from their lips. Two ponies and a horse, a soft bay with movie starlet lashes, blinked out across the river at the Manhattan skyline.

"Well, you found them. They're all yours," the woman said, pushing the handle of the pitchfork in my direction. "If, that is, you can pick a stall."

So began my season with George and Ann Blair, or Dr. and Mrs. Blair, as I was instructed to call them. They were the proprietors of the New York City Riding Academy, the stables that shared the edge of Randall's Island with the towers of the Manhattan Psychiatric Center. Dr. Blair was a retired deputy chancellor of the New York State University system, professor-turned-urban-rancher and, like his wife, an African American from Queens. He had a shock of snowy beard offset by his dark skin, dusted with freckles that he kept shaded under a cowboy hat at all times. The couple were creaky with age, and even the upkeep of just three horses was almost beyond their capacities, though Mrs. Blair tried.

Mrs. Blair is to date the most meticulous stall cleaner I have ever met. Her perfectionism was compounded by her age, so that it took her an entire day to do the three stalls, sifting through each shaving for every last speck of horse leavings the way a prospector pans for gold. The only plus of her obsession was that in the end, she turned out to be far too compulsive to ever let me muck. I simply could not meet her standards. Shortly after becoming the couple's helper, I was summarily banished from cleaning the little red barn.

With a backdrop of skyscrapers, the cacophony of traffic on the river's far shore seemed muted, but maybe it was simply the effect of how arresting the horses were in a setting that seemed impossible. On the island, the speeding cars fell as little more than the shush, shush, shush of ocean waves. My only duty was to ride.

Mrs. Blair had vetted my riding the morning I arrived to explore. After we picked out the barn, she had me tack up her portly mare with the Greta Garbo eyes, Brownie. I was wearing bike shorts and sneakers. She led the mare into a paddock framed by a chain-link fence and crossed her arms. Surely I wasn't supposed to get on in this getup?

"So?" she said, turning to me. "Ride."

I snapped my bike helmet on my head, crammed my sneaker in the stirrup iron, and swung up on the squat mare. Then I took a gamble: I'd spent all of a half hour with Mrs. Blair and already could tell she was the kind of horsewoman who is allergic to anything other than straight talk and action. They're a sun-baked kind you find everywhere in the horse world, whether in the stone-dust arenas of show barns or the scrubland of ranches. They are a type steeped in the silence of their equine charges, who view the human need to communicate with words—and fill the quiet moments with chatter—a failing of our species.

There atop Brownie, I realized that this wasn't to be a dressage test, where I would impress Mrs. Blair with my subtlety, caressing the animal invisibly so that it danced below my bike shorts. And this wasn't even to be a show ring school, where I'd prove my mettle by diligently exercising the horse's every muscle, slowly ramping up the workout with bursts of interval training. And it wasn't a trick, like my parks department test on Samson.

In the thin line of Mrs. Blair's lips, pursed to the point of disappearing, I saw this was to be a different sort of trial. *Show me what you got*, her silence said in a thick New York accent of its own. What else had I expected at an NYC barn under the gritty shadow of a psychiatric ward? She threw me astride Brownie in my shorts, I realized, to answer just one question: Was I cowboy enough?

I took a breath and jammed my sneakers into Brownie's sides.

Brownie was a quarter horse like Amigo, I could tell on sight, and thus possessed of a Labrador-like loyalty that manifests in obeisance to a fault. A quarter horse will tip off a rock ledge and straight into a ravine if the rider wills it.

Brownie was born game. Like her kin, she had ham-hock haunches jammed with muscle, and though they were flat and unfit—the Blairs never rode her—they still bulged with latent power. Quarter horses come fully loaded with prodigious hind ends designed for blasting after a wayward calf. On a dime, a horse like her can skid heels deep into the dirt to brace against the might of a steer thrashing at the end of a lasso.

In the middle of my city, we were far from ranchland and grazing Angus, but I could nevertheless tell from her rippled haunch and wise eye that Brownie retained a ranch horse's giddy-up. And so at my crude kick, she lit out, a 0-to-60 hop that belied the little horse's potbelly. We lapped the paddock at a straight gallop, my bare inner thighs burning against the stirrup leathers. On the far side of the pen was the skyline—squat brown buildings and wooden water towers and teeming streets. I turned my head as we cantered past it all and watched it become a smear of city, an uncanny view astride a soft brown horse.

We rounded the fence line and circled past Mrs. Blair, where she stood, poker-faced and with her arms crossed, at the gate. I flicked Brownie's reins. It was a gesture the Hess brothers had "learned me," as they might say, at Deep Hollow. It was an infinitesimal tug on a totally slack Western rein, so different from the taught connection that creates the dance of the English discipline I typically practice. In the *haute école* of the Western discipline, it feels to me that the rider does not tell the horse what to do, but merely opens a door for the horse to step through. The touch on the reins was an invitation to Brownie, not a command—the tap on a shoulder that brings a daydreamer back to earth.

The well-broke mare took my cue and stopped dead. I was jarred in the saddle as she braked hard. I straightened my bike helmet from the jolt, and patted her neck. Mrs. Blair was still silent. "Now imagine what I could do if only I had britches on," I said. At last Mrs. Blair cracked. She laughed.

I peddled back over the footbridge that day with blisters on my inner thighs, rubbed raw from the leather. And a job working for the Blairs.

Dr. Blair started the New York City Black World Championship Rodeo in 1984, an extravaganza of bull riding, calf roping, and bronc busting. The festival pitched up on the pavement each year on Lenox Avenue at 145th Street in Harlem. It had its own official day of the year, proclaimed annually by the mayor.

"John Wayne is the epitome of the Western cowboy, and there was no clue that Afro-Americans were even cowboys, that they're part of the American culture," Dr. Blair told me during a chat in 2019. He was eighty eight years old and recovering from a broken femur, sustained when a horse acted up as he loaded it on a trailer. He was healing up for still another summer on the island. "That's why I started the rodeo," he said. "Because the black cowboys were erased?" I asked him. He snorted. "They haven't been erased because they were never included."

By the time I started working for the Blairs, the rodeo had wound down, and all that was left was their tiny riding school on Randall's Island, on land they leased from the city. It was dedicated to educating poor urban youth about horses, cowboys, and their place in American history. But when I started the job, I didn't know about the riding school's mission or the by-then-defunct, pioneering all-black rodeo Dr. Blair had founded, or why. Back then, I thought it was a just a pony camp run by two lovable, ornery old folk.

My job consisted of exercising Brownie and her pony peers, since the Blairs were both past the age of riding. But my main duty was performing riding demonstrations for the public school groups that visited the red barn. Part of Mrs. Blair's silence, I was to learn, was her playing yin to the yang of Dr. Blair's volcanic fury. Most days he sat in a folding chair beside the little red barn while his wife picked stalls endlessly, reading the paper, twirling his white beard, and barking orders. I ignored it mostly; I groomed Brownie or contemplated the skyscrapers from atop her back while Mrs. Blair panned for precious metals in the horses' stalls, and I learned, like she had learned, to endure it. But when he turned it on the schoolchildren, I cowered alongside them.

"Be good children, and do everything adults say or there will be consequences!" Dr. Blair bellowed, startling Brownie, whose reins he held as I sat mounted on her back. I had been waiting to demo a few moves before the third graders, now growing wide-eyed as they stumbled back from his roar on the grass. They were in awe of him and his molten rage and white beard under his cowboy hat, but most of all in utter disbelief that Brownie was real. Dr. Blair invited only the most underserved children of the city, from the poorest tracts of places in the Bronx and Brooklyn, to visit his island equine paradise. When he asked who had ever seen a real live horse before, not a single hand ever went up.

What perplexed me most was that his speeches never had anything to do with horses; in fact, mostly we didn't even teach our tiny guests to groom or ride. Sitting on a fence post, as each class filed off their yellow school buses, he raged at them to grow up to be good men and women, to be safe, to stay away from guns and gangs, and to finish school, and to stop fidgeting at once! As resident cowgirl, I was a magician's assistant, demonstrating the trot and canter on Brownie between his tirades as the children tried desperately to cork their awe and their youth and sit still.

"Dr. Blair, sir?" I said one afternoon as I doused Brownie with water from a garden hose beside his permanent folding-chair throne. He looked at me with bright eyes from under the brim of his white Stetson. I had never addressed him directly in the months that I worked there, following instead the head-down-and-do-chores strategy of his wife. He put down his newspaper and bored his eyes more deeply into me as a response.

I had wanted to ask him something since the first class. I had held my tongue all summer ever since that initial group of children filed back onto the bus. They were deliriously happy but had gotten no closer to riding a horse other than to stroke Brownie's nose as she nuzzled their sea of outstretched hands like she was a pop star high-fiving fans in the front row. I struggled with how to phrase my question. I wanted to ask, "Why don't we teach riding lessons here, at a place ostensibly called 'New York City Riding Academy'?" On the three horses, it would have taken us a day to mount up the forty-plus kids who came at a time, an impossibility, of course, but weren't the children coming to ride?

Every session I was washed with extreme guilt, a well-off white woman prancing on horseback before a sea of mostly impoverished, majority-black children. I was plagued by the feeling that as I demonstrated how to ask a horse to back up, to trot, and to canter, what I was really demonstrating was my privilege.

"Dr. Blair," I finally said. "What's the point of our classes?" I frightened myself with my bluntness and busied my hands with untangling the hose. Out of the corner of my eye, I saw Mrs. Blair pop up over a Dutch stall door and cock an ear.

"Sarah," he said. "Do you know what a cowboy is?" The hat was off, and he raked his hands through what remained of his white curls. I thought I did. Doesn't everyone? So deeply entrenched in the American psyche is the cowboy that at first I thought it was a rhetorical question.

He stood up from the folding chair. "A cowboy is a black man."

The cowboy kicked the shaggy bay gelding into high gear. It was a cold day just before New Year's Eve 2018, and I was far from that little island and that old memory in Harlem, in a pocket of Texas called Rosenberg, about twenty-five miles outside Houston. I was inside a ramshackle barn in a half-covered riding ring. A dozen Black-Mouth Curs, cow dogs, lounged in wire kennels beside the arena. They bayed at the sight of the man on the horse; they knew a cowboy when they saw one and keened their wish to be out with him, punching cattle.

Larry Callies, the cowboy, was exercising the big bay before an audience of the dogs and a pen of yearling steer who were rubbing their horns idly on the fencing. I watched too, alongside the little Longhorns, as I sat on a chestnut quarter horse named Bud. Callies was in the market for a new ride, and we had driven out to this ranch on the outskirts of Rosenberg to test-drive a chunky gelding named Swamper.

With a swish, Callies loped into a corner of the ring and rocked his mount back on its haunches just by a flick of his wrist. The animal stopped expertly, dead square, then spun on a dime. The dogs yelped. Even the calves seemed impressed.

Swamper was not a looker, a fact Callies had been loudly reminding the horse about all morning. The gelding had two beady eyes gone smoky under a film of cataract and a sway-back that begged for a saddle to hide it. But he was pretty spry for any horse, much less one who was pushing twenty-two years old.

"I'm going to offer half for this horse," Callies said conspiratorially to me as he rode Swamper alongside Bud, looking around to be sure he was out of earshot of the rancher who was selling Swamper. The seller was preoccupied; he was in a round pen at the arena's far corner, riding a shivery buckskin colt. His eyes were half-closed with the intensity of concentration as he coaxed the barely broke colt to a lope and stuttered him down to a jog. But there wasn't much chance Callies could be heard: his voice has creaked like a barn door ever since the nineties, when the cowboy's vocal chords mysteriously decided to stop working. The malady derailed a burgeoning career as a country music crooner. Callies became a postman. "It is a fate about which Mr. Callies is relentlessly upbeat," I wrote in an article about him in the *Times*, "smiling his wide newscaster smile as he explains that if he had ended up a country music star, he would have had less time for his true passion: rodeo."

I leaned out of the saddle and stage-whispered back: "So

you like him, then? Even though you can't stop saying in front of the poor thing how ugly he is?" I said. "You know, Swamper has ears, Larry."

"I don't like him," Callies said, the rust rattling in his throat. He pulled off his dark felt cowboy hat and pressed it to the side of his face so it shielded his mouth from any eavesdroppers, steer or human. "I *LOVE* him. I could rope any steer in Texas off the back of this here Swamper. But don't you dare breathe a word about it. I'm trying to get him for half-price!"

We cackled conspiratorially and both dismounted, careful to avoid a lone, foolhardy Black Mouth who had escaped the kennel. The working dog slalomed in between our horses' hooves heedless, whining to be pet like any lapdog. I tied Bud to a hitching post made of a horseshoe nailed into a wood beam, uncinched his rope girth, rubbed him down, and plaited his long Western pleasure horse tail, the cur nuzzling my pockets all the while.

In his rodeo shirt, Wrangler jeans, spit-shined boots, and black felt Stetson, Callies, age sixty-six, groomed Swamper beside me, every inch the cowboy. In case you missed the memo, the bumper sticker on the pickup we drove there said, "Gone Roping," and the flatbed was full of nothing but rawhide and lassos. Callies was a cowboy to the caterwauling cow dogs, and pure cowboy in the saddle, implacable as the old but catty Swamper had jigged and zagged around the ranch, Callies's brow furrowed in concentration, feeling the animal out.

"But not everyone sees a cowboy when they look at Mr. Callies," my *Times* article continued. "Racism and history's omissions have meant that for many he's miscast: Mr. Callies is black."

Historians and Hollywood have erased black cowboys from their rightful place among the sunset and sagebrush and the Western vistas of our minds. Prejudice to this day has continued to finish the job, painting all-white portraits of the American frontier. In fact, *one in four cowboys* on the frontier were black, according to data and historical accounts from the pioneer era, which began in 1865 after the Civil War and ended in about 1895.

Yet their legacy has been whitewashed from Western iconography and American lore by John Wayne and Sergio Leone and written in invisible ink in the pages of the Lone Ranger comic books.

It is an ugly omission, theirs a true story that society has long neglected to tell, bias and bigotry putting only white skin under silver-screen Stetsons. Blacks were lawmen meting out frontier justice, like Bass Reeves, an Arkansas man who self-emancipated from enslavement and was commissioned as a US deputy marshal in 1875. (Some believe Reeves was the inspiration for the Lone Ranger himself.) They were cowboys who were as famous in their time as Roy Rogers, such as Bill Pickett, the inventor of the rodeo sport of bulldogging, rassling a calf to the ground with bare hands. Though he was born in 1870, he was inducted into the Pro Rodeo Hall of Fame only in 1989. He was its first black honoree. And they were outlaws, as notorious in their day as Billy the Kid, like Cherokee Bill, who used his race as an asset for the gangs he joined: his skin meant free passage through Indian-controlled territories forbidden to white renegades and staying one step ahead of the law. But I bet you've only ever heard of one of those Bills.

Some believe that the word *cowboy* itself points to the blackness of those who bore the label. First recorded in the English language in the early 1700s, initially it referred to *actual* boys, that is, kids who tend to cows. By the 1800s, the definition had changed to the term we know today. And yet during that same era, "boy" was often a pejorative reserved for blacks. "A cowboy is a black man," Dr. Blair had said.

I was in Rosenberg that day searching not for horses but for the black cowboy's story. I found it tucked inside a former barbershop at the edge of town. Today it is a tiny storefront museum that Callies spent his life savings to found: the Black Cowboy Museum. He opened it in 2017, his shrine to cowboys who are enshrined few places else.

Until he was an adult, Callies didn't know what role people who looked like him play in the country's story, no conception of the black cowboys' rich legacy. Then on a rainy afternoon nearly twenty years ago, he was clearing out a cluttered tack room at a guest ranch where he was employed as a cowboy, when he came across a photo on its way to the trash. It was from the 1880s and showed eight cowboys and eight horses.

Seven of the cowboys are black.

Looking into those faces, he saw faces like his own for the first time as part of the narrative of America's invention. In the black cowboy's erasure from the story of the West, Callies saw his family's own marginalization from the annals of cowboydom. He counts among his extended family some of the most talented bareback bronc riders and calf ropers in the state's history. But few of their names are etched into any rodeo buckles or lauded in rodeo halls of fame. Until the 1970s, most modern rodeos were segregated. Blacks rode in rodeos of their own, where whites were able to compete, but the same was not true vice versa. Unable to saddle up in sanctioned events, blacks had no shot at state titles, championships, and official cowboy glory.

That antique photo of the seven black men greeted me as I entered the lobby of his storefront museum, where Callies welcomed me that same day we tried Swamper, turning from postman to docent as we crossed the threshold. Callies opened his museum to stake his claim to his heritage. He had had enough of people doubting who he truly is: the son of a cowboy, who is the son of a cowboy, who is the son of a cowboy.

"I used to go to school and people would kick me if I had a pair of cowboy boots on," he told me as we wandered the museum, a motley collection of old saddles and rusty six-shooters. As a little dude in El Campo, Texas, in the 1960s, he grew up learning roping from his father at the family's ranch in a still-segregated town and country. "I had to quit wearing them because they would beat me up. Whites would beat me, told me the boots were not my own. Blacks would beat me, told me I was a black guy who wants to be white."

Callies's sandpaper voice crackled, this time not from the affliction of his vocal chords but from the pain of the memories. "This is who I am. I was always going to be a cowboy. If God made me white, I was going to be a cowboy. But God made me black. And I am a black cowboy."

The poet Langston Hughes wrote in 1942 in his poem "Merry Go-Round":

Where is the Jim Crow section
On this merry-go-round,

Mister, cause I want to ride?
On the bus we're put in the back—
But there ain't no back
To a merry-go-round!
Where's the horse
For a kid that's black?

Twenty-five years after writing "Merry Go-Round," Hughes called William Loren Katz on his rotary phone. It rang in Katz's apartment on Sixteenth Street in Manhattan. The year was 1967. " 'Don't leave out the cowboys,' " Katz recalled the poet saying, when we spoke recently. "He said it twice."

A Jewish man from New York City, Katz was one of the foremost scholars on history's omissions—adding back the women and people of color omitted from so many narratives by the white and male authors of the historical record. At the time of the fateful and unexpected call, Katz was writing a book on African American history. Hughes had heard about Katz's project, and urged him to dive deep into the cowboys missing from the range. "He felt for black children and white children to see black people as part of America, they had to learn they were part of this Western movement, this pioneering, frontier activity that is so much part of the American story," Katz recalled.

"*Don't leave out the cowboys*" has become Katz's life's work.

"No picture of American history has been painted more white than the pioneer picture, the story of the frontier. It's more American than anything else and it's been represented that way, as totally white, in movies and countless novels—and that had to go," Katz, ninety-two years old when we chatted, told me. "Because it just wasn't true." From time to time, Katz and Dr. Blair had worked together over their decades bent to their shared cause of inking back characters into history's story. Katz's pen was indefatigable; he wrote forty books, and when we spoke shortly before he died in the fall of 2019, he was still at it.

His work expanded for me the story Dr. Blair began that afternoon in 2004. Not only were a quarter of pioneering cowboys black, as the seven faces in the photo Callies saved laid bare, but the West was integrated—at least far more so than the eastern states the pioneers had left. In some territories, whites initially tried to establish the divisions that ruled on the other side of the Appalachian Mountains, but in the raw West, where survival depended on cooperation, segregation didn't work. People needed one another, Katz told me, whatever their color. Out on the western front, among the cowboys sharing cups of billycan coffee over a campfire, was something closer to equality.

"For African Americans, even more was riding on their march west," Katz wrote in his book *The Black West*. "They carried a greater need, a heavier burden, and paid a higher price. More than Europeans, pioneers of color pined for a home of their own, a place to educate children, protect women, and nail down elusive dreams."

The modern force that whitewashed the West and threw the black cowboys from the tack is hatred at its most insidious. Black cowboys were erased because to the authors of our country's myths about itself, cowboys—and all the glory, honor, and Americanness they stood for—couldn't be black.

"The West was part of the mythology of America; the cowboys, the narrative of the pioneer spirit represented the best of us," Katz told me. "And if the black people came along, they came along at the end of a whip and in chains. And that's not the American tradition people wanted to remember."

In Rosenberg, the museum tour complete, Callies and I pulled up folding chairs inside his little gallery and talked horses, swapping phone pictures of our favorites. Beside us was a display of posters from novelty films he had collected, like 1925's *Chocolate Cowboy*, where the star was a white man in blackface. In *Harlem on the Prairie*, the black cowboys themselves were viewed as the punch line. In a 1937 review of the film in *Time*, the reviewer is incredulous at any blackness in the West: "In this apocalyptic land everybody—the prospectors and stagecoach drivers, the medicine men, outlaws, sheriff, the hero with the silver-plated stock saddle—is a gentleman of color. No attempt is made to explain how so much pigment got all over the open spaces. In Negro theatres it will be a conventional Western, and it can play the artier white houses as a parody."

On a far wall, leather chaps embellished with red hearts were carefully displayed in a vitrine. They belonged to Callies's cousin Tex Williams, who, Callies believes, was the first black man to make it to the National High School Rodeo Finals. Light-skinned, Tex slipped in among the teenaged competitors that day in 1967. The stallion launched from the chute with the force of a dust storm, and Tex sat every whipsawing buck on that raging bronco, Callies recalled. The stadium erupted in cheers—for exactly three seconds. That's when the horse bucked Tex's cowboy hat clean off, revealing what Callies described as his cousin's "nappy hair."

"They booed," Callies recalled. "They booed as he rode that winning ride."

Next to the photos of Tex on his many electric rides and the other exhibits, I noticed that in place of any scholarly museum plaques were haphazard magazine clippings explaining the black cowboy's place in the world or articles printed off the Internet. The whole place felt less like a gallery and more like a family's attic—here an uncle's old spurs, there a cousin's collection of his trophy buckles—festooned with pictures of relatives yellowed with time.

But I couldn't begrudge the little museum its lack of rigor. Callies is a mailman, not a historian; a cowboy, not a curator. He had taken his history into his own hands. And just as Swamper was an overlooked horse in which Callies could see his beauty, here was an overlooked story that Callies was demanding the world see too.

"I didn't want these cowboys to have lived in vain," Callies said, pointing to the antique photo of the seven black men who had started it all. He swung out his arm to take in the family albums that filled the museum walls, of cowboys he loved kept out of the rodeo arenas and halls of fame in which he believed they belonged. As more people heard about his mission, they came to him with saddles and spurs, etchings and daguerreotypes of black cowboys to donate to his collection. "This is why I opened this museum. Black cowboys meant something," Callies said. "And they meant something to me."

There, on the chilly December day in Texas, I recalled that hot August afternoon in New York City more than a decade back with Dr. Blair—that day he taught me about black cowboys. The day I stood holding Brownie and asked Dr. Blair why we didn't teach riding to the children who came to see us on the little island between Harlem and Queens, why we taught only history instead.

"These children need to know there are other lives to live; other lives than the ones they know, than the ones that seem to them inevitable, stretched out before them in the housing projects and in the streets," Dr. Blair said then. "These children need to know that they are part of this country's story." As the cars swished by along the FDR Drive, their sound swept across the water, in time to Brownie's tail.

"I am not teaching them to ride," Dr. Blair said. "I'm teaching them to dream."

ADONIS

One day during summer break from college, I went to visit an old lease horse of mine. Reggie. His show name was Instant Karma, and I recall he was an absolutely snow-white, eighteen-hand-high warmblood with cream-colored feet the size of a Frisbee, a pale pink muzzle, and eyes rimmed in dots of black. As he thundered around the ring at a barn not far from my home in East Hampton with his new lessor, a strange little horse clip-clopped nearby. She had what is called a dished face, concave at the bridge of the nose, with nostrils that trumpeted out like a honeysuckle. It was stuck on the top of a neck she carried stick straight in the air, like a lollipop.

I had rarely seen Arabian horses in the Hamptons, that place where Jackie Kennedy rode bred-to-death thoroughbreds across the potato fields in custom tweed. Or in the jumper world in general, where glamorously lugubrious European stock is now the ride of choice and Arabians are generally derided as useless, diminutively sized with tart personalities and weak, flat jumps. Arabians are stately steeds in their own right—delicate, effervescent, and haughty. But they are out of place in the equestrian world of the Hamptons like a Maserati is at school pickup.

Like babies, "all horses are beautiful" is something I know I'm supposed to say. This one wasn't. She was an Arabian mutt. And ugly.

"Where'd you get . . . that?" I asked the tween riding, or trying to, jangling against her mare's spiking trot as I leaned over the fence.

"A lady from Tennessee sent her to me. Free," she said, pulling up on her sweaty horse, her shrug confirming that she knew what she sat on was homely, but a horse all the same. And the implied: Who among us would say no to a horse?

And then, as if it made all the sense in the world, the girl continued: "The lady's got heartbreak." She kicked the sour-faced mare back to a trot. "And a lot more horses to give away."

A few more equally baffling exchanges later, and I was in possession of the lady's phone number. My college tuition had put horses out of reach for my family. Before I left for college, I had sold Willow to an elderly woman who doted on her until the mare died at age thirty-two. In the intervening years, I had leased a few horses like Reggie during school vacation, a few months at a time, with money saved up from my summer jobs. But buying a horse of my own was out of the question. A free one was another story. Horseless and curious, I called the woman from the barn parking lot.

She picked up on the first ring, as if she had been expecting me.

So began an hours-long unburdening of her soul to a complete stranger as I sat held captive in the dusty barn parking lot by the conversation. I wasn't a journalist yet, and I've since become used to being a vessel for trauma. It pours through the phone, those moments when my anonymity to the person on the other line becomes for them the lattice screen of a confessional booth. But at that moment, I was overwhelmed by the lady with the raspy voice from Tennessee telling me her story, a grown-up saga from the soap opera of adulthood to which I had not yet in my young life tuned in: a splintering marriage, a car crash, and an equine custody battle.

I hung on listening, however, because in between the sobs, she kept promising me a horse.

Her name was Juliet Faust. Juliet and her ex-husband had founded the Tennessee Valley Hunt in the Smoky Mountains back in the eighties. They had met and married in New York City, but she followed him to the woods on the promise he would build her a horse farm. He did, and for thirty years of marriage they raised horses as their children, interlinked generations of animals who were all extended family members of one another and the Fausts.

In truth, her husband did not love horses, she believed. He loved whiskey. His other love was for micromanaging Juliet, and his constant complaints about the quality of her farmwork, the fitting of a horseshoe, or her choice of a vet led her to cede barn operations to him.

For Juliet's husband, hunting was the perfect marriage of all of his pursuits: booze, horses, and cultivating an air of lordly command. Each hunt begins with a traditional sip of what is called the stirrup cup, a circa 1700s tradition of emboldening riders with a glug of port punch. The day's chase is always followed by a boozy hunt breakfast. (By the way, it's called "breakfast" even if it happens at midnight. And don't you dare call a hunt *hound* a "dog." Hunters love arcane decorum almost as much as they love brandy.) Then there are the mid-gallop nips of warming drams of scotch from flasks secreted in the top of a boot or from crystal flagons slung from the saddle.

In an article on the Fausts' Valley Hunt in the *Metro Pulse*, a Knoxville, Tennessee, paper, the reporter described the group the couple founded much as one might describe hunts the world over: "A cocktail party on horseback. They pop champagne bottles and pour glasses for everyone. Several riders sip from stainless steel flasks. A few slur their words." One of those wobbly riders interviewed was Juliet's husband, who responded to an inquiry about the sufficiency of the champagne supply thusly: "Just perfect. Nobody's fallen off their horse—yet."

But fall off he did, Juliet said, five separate times, breaking bones with every tumble. But what ultimately got him was not a horse: "One night, he crashed his car into the side of a mountain," Juliet told me in the barn parking lot in the first minutes we ever spoke. "He broke his neck, and I cared for him for months, as he was unable to clean himself, feed himself, or do much of anything. I did it all."

The night he rammed their Suburban full of saddles into a mountain on his return from kenneling the hunt hounds was the last time he'd break his neck. That is, he survived, but the bones set poorly. "The doctor told him he could never fall off again," Juliet said. Given his predilections, she added, "Naturally, that meant he could never ride again." After thirty years of riding side by side in hunts all over the world, horses were the only thing left holding them together. Unable to ride, they were unable to last.

I still remember what she said next, word for word, from when I spoke to her in the Amagansett parking lot over a decade ago: "When he recovered after the car accident, he looked up at me and said: 'I saw my life flash before my eyes, and it was worse because you were in it.'" I also remember my gasp across the years: her tears, the smell of the sun-heated seats of my mother's silver SUV.

"That's why I have to get rid of all our horses before he's back on his feet," she told me back then, a stunned stranger, twenty-one years old, and overwhelmed as I listened to another woman's life unravel. "They were our children, and he does not deserve them." There were nearly a dozen horses, she explained, a herd of mothers and brothers, babies and fathers. Juliet's beloved brood.

I sat silent, watching a trapped horsefly ping against the windshield. "I'm so, so sorry," I said after a time.

"When do you want your horse?" she asked.

Juliet sent a trailer with Tennessee plates to East Hampton. Two days after that first phone call, it pulled into the circular driveway at a friend's backyard barn in East Hampton. The cab opened, and a mustachioed driver chewing tobacco stepped out and spit on the meticulously kept flagstone off the drive. "Anyone want to sign for this horse?" I grabbed the pen and scrawled.

The man strode charley-horsed steps up the ramp and clattered chains, swinging a netted bag full of hay out of the way as he led down his charge into the bright Long Island morning. I braced for a sway-backed, dish-faced Arabian mutt, ready to fiercely love whatever tumbled down the ramp. Out into the sunlight strode a stunning six-year-old chocolate bay warm-blood. His shoulder blades towered seventeen hands high. He stepped down the truck ramp with a parabolic stride as lithe as swaying beach grass. His name was Darius, according to a sheaf of papers the driver thrust into my hands, along with the end of a lead rope attached to the jaw-dropping creature.

"You're an Adonis!" I exclaimed at the other end of the tether. It was his name forever after.

For as long as I had Adonis, I never met Juliet.

घोड़ा

[Ghōṛā] *Hindi:* horse

Through the heart-shaped space between the curlicue ears of my stallion, I looked out across a stone quarry on the outskirts of Udaipur, India. His fluted ears bent toward each other until their tips touched at the top of his head, a paragon of his rare breed, a Marwari. Unique to the indigenous Indian animal, the ears were like two question marks that met in the middle to form a perfect heart of space between them. Through it stretched the green marble of Rajasthan.

It was 2007, and I was on my final journey in the pinch-me period of my life when I was—get this—a spa reporter.

My journey to that cushy job had begun a year earlier in New York City, with my typical modus operandi: failure.

Days before graduating from Columbia University, an old colleague at the *East Hampton Star* stunned me with the opportunity to interview for a job at *Vogue*. I made it past thirty-two other candidates, I was told, and through five rounds of interviews, until I sat before the publisher. It was the final interview before I'd meet Anna Wintour, high above Times Square inside Condé Nast headquarters. Wordlessly, the publisher pointed a lacquered nail at my résumé: a typo.

"I'm not perfect," I said, flushing with that old Brearley shame.

She pursed her lips. "*Vogue* is," she said. Then she told me I had a button undone on my blouse.

I fled out onto Forty-Second Street, tears streaming, when I suddenly realized I was crying with relief. It was years before *The Devil Wears Prada*, and I already knew that nobody turns down a job at *Vogue*. As I had made my way through each round of interviews, I felt a future I wasn't sure I wanted—I loved fashion, but did I want to *live* it?—barreling down on me with the inevitability of a horse on a longe. But *Vogue* had turned *me* down. And the world was waiting.

On the sidewalk in front of 151 West Forty-Second Street, I called Delta and booked a one-way ticket to England.

In London, I lived off money I'd saved mucking stalls and leading trail rides while I pitched story ideas madly, cold-calling any and all publications. Work was piecemeal and paid little. Then I hit the jackpot: an editor at the *Times* of London had a story that needed writing. A public relations company had offered to fly reporters to visit a client's luxury hotels. Such trips are called junkets and lavished on journalists in exchange for a write-up. No one from the newspaper could make the next trip, and it was leaving tomorrow, she said. Oh and the free vacation would be to review spas. Was I busy?

Was I busy?!

So began a year of my life being sent all over the planet to be scrubbed with spices and sloughed with charcoal, soaked in goat milk and lubed with yogurt, pumiced, peeled, and prodded, all in the name of journalism. And all my travels, from Mexico to Greece, from Egypt to the Arctic, were on a public relation company's dime.

The *New York Times*, where I have worked for the past decade, admirably pays its own way, always. It forbids reporters accepting such freebies, viewing them as tantamount to a bribe for coverage. But moral high ground is a luxury that only a dwindling number of legacy publications can afford to stand on. And so long before I joined the *Times*—and never after—I made my living writing for publications that permit such trade-offs, soaring around on press junkets. And on every junket, I sought out a moment to ride.

I couldn't always find horses: it was donkeys in Morocco's Atlas Mountains, reindeer in arctic Finland, a camel named Cleo beside the Nile.

That ride in Rajasthan was on my last junket, because right before I left for it on another trip, I had an epiphany. It hit me as I lay in a cool cave deep in the volcanic rock of the island of Santorini in Greece, full of spanakopita and steeping in a wrap of red wine and honey. In the dark, I reveled in the beauty treatment I was being paid to review, marveling at just how awesome my life was. Then, thundering into the dim cave came another thought, echoing across two thousand years of time. It was an ancient Jewish proverb, and I heard it in my head and in my bones in ancient Hebrew: "If I am not for me, who will be for me?" They are the words of Rabbi Hillel, a Babylonian sage born in 110 BCE.

There in the cave, I was doing a pretty good job of being "for me," as Hillel instructed—in fact, so much so that I had been living a life of extreme self-care, wholly devoted to my own pleasure. But that phrase is only part of Hillel's wisdom. His next sentence burned bright in the cave: "But if I am only for myself, what am I?"

I shot up in the dark, a cucumber slice falling from my eye. My life was rich, resplendent, so pleasant it bordered on ambrosial. I was not proud of myself. I had no cause to be. My father and mother devoted themselves to others, to helping, to fixing. They had taught me Hillel's words, made them a trope of my upbringing. They lived by the Jewish principle of *tikun olam*, a worldview that says acts of loving-kindness can repair the world. I had loving-kindness in spades—all of it directed at myself. And all it had repaired was sun damage.

I peeled the wine-soaked strips of linen off me, unwrapping myself like a mummy. Apt, as I felt wakened from the dead. There on Santorini, I changed my life, repositioning my journalism to the good of all, not just the good of Sarah. Since that afternoon, I've worked toward Hillel's ideals, forging a career in which I aspire to give voice to the voiceless and shine light on the darkest of lies. At that moment in the dimness, I had to start immediately, so I jumped off the table and shot out the mouth of the cave, because Rabbi Hillel's final sentence echoed after me: "And if not now, when?"

As soon as I returned from the Greek junket, I applied to Columbia University's School of Journalism, and though I still can't tell the difference between "their" and "there" (or is it "they're"?) I got in. In 2008, I headed there (Did I do it right?) to pursue a master's degree that fall, the initial step in Hillel-ifying my journalistic life.

But first, to Rajasthan for just one last spa . . .

*   *   *

I watched from the tiny plane as the northwestern Indian city of Udaipur came into view, wrapped around the saucer of an artificial lake. The water of Lake Pichola rippled below as we passed the maharana's palace on its shore. Fateh Prakash Palace houses a collection of English crystal, the obsession of Maharana Sajjan Singh, the castle's nineteenth-century proprietor. The crystal is formed into every sundry item you can hew out of rock, and some you probably shouldn't: knives, candlesticks, thrones, even a very glittery but very uncomfortable sofa.

All the sparkles were shipped from the manufacturer, F. & C. Osler in Birmingham in the United Kingdom, across the Indian terrain. For portions of the more than four-thousand-mile route, some of the homewares rode strapped to horses like the Marwari, and the stouter Kathiawari, horses. That caravan arrived too late, the story goes—the maharana died before he ever snoozed on his crystal couch.

I spent the week in a spa on the edge of that lake, including reviewing a cumin scrub that made me smell like a rotisserie chicken. I spent my mornings being marinated; every afternoon I asked the concierge the question I ask all locals when I'm on a reporting trip: Where can I find the horses?

"They'll never let you ride, you know," my chauffeur said to me after a few fruitless days, peering back over the headrest of his white Lincoln Navigator as it rumbled down a rutted road. True to the excesses of junket living, every reporter on the trip had a personal driver. He was right: on press junkets, journalists are more carefully handled than all that crystal. I was passed with kid gloves from one minder to the next, an insurance policy so I would have nothing but an exquisite experience to report to readers. "They're worried the review might not get published if you get bucked off and die," he said, laughing.

"But," he continued, "I know a guy. He's got a *ghōṛā*," he said in Hindi. I looked at him quizzically. "A horse."

The next day we set out for a sightseeing trip the publicists had arranged, to a miniaturist painting gallery to watch artisans doodle with the finial hairs of a squirrel's tail. But as soon as we were out of sight of the PR minders, my driver gunned the Navigator in the opposite direction.

We pulled up to a hillock, where two men stood at the top of a gravel drive leading to a marble quarry. One of the men stood out: he wore a red beret over his white handlebar mustache and a khaki top festooned with epaulets. He had on classical jodhpurs of a sort I had seen only in antique hunt prints: they ballooned loose at the thigh and tapered into his dusty field boots. The man, in the full uniform of the Indian cavalry, leaned against a gnarled Khejri tree.

Beside the men were two of the most uncanny horses I had ever seen. The Marwari is a horse native to India. The pair of animals, one black, one mottled, white with a cup of café au lait spilled across the hide, held themselves erect like Arabians. They were massive like thoroughbreds, undulating muscle beneath threadbare skin. They were well kept but lean, the faintest trace of rib showing as they sucked in air, skin taut on their bodies like the pelts of racing greyhounds. They stamped at the horseflies settling on their skin and peered intently at the sparkling white Lincoln from heads held high like a pair of ibexes.

And their ears.

For the most part, Marwari horses look like many other horses and come in different grades. Some are thicker bodied— short from sternum to tail, with chiseled heads. Others are finer creatures, every inch of them arched, sloping, willowy, and tall. Some are speckled; others are pure brown, red or white, or splashed with calico color. Generally Marwaris are midsized horses, but sometimes they're small. A few are massive. Often they are sleek, their heads placed high above a strong shoulder, wise of eye. Others are plain, thick, and common. Spread across the subcontinent, the Marwari phenotype changes depending on where you find them around the country. All that pheno-typic variation is likely a result of Indian horses becoming war brides to the animals brought by India's many invaders over the centuries, from the Mongolians to the British.

But they all have curlicue ears. No one knows exactly why, but even if they did, here's where the reporter in me occasion-ally relents to the horse girl who also rents space in my soul: I really don't want to know why. Just like I'd rather imagine a million fantastical reasons for the rainbow that occasionally arches over the barn rather than the pedestrian reality of prisms and refracted light, I want to believe Marwari ears are magic.

Everything curled is better: pug tails and pigtails and roller coasters and curly fries. Marwari ears take the beauty of the horse and top it with a sundae swirl of whipped cream. And everything is improved with whipped cream. It helps that what lies beneath those funny ears is bold and kind, possessed of incredible endurance and clever enough to command Indian dressage. In it, Marwaris piaffe, or dance in place, as beautifully as any Lipizzaner stallion trained in the *haute école*. But at that moment, all I could see was a lily, gilded by two adorable, fuzzy, curlicue ears.

My guide introduced me to the men. One owned the horses; the other was a cavalry officer who was to be my guide. The officer looked at me skeptically. I hadn't brought riding clothes to India and was decked in silver *jutti* slippers and white linen pajama pants I'd bought in a bazaar in an effort to have something other than shorts and flip-flops to wear. I knew his askance look. It was the same one I gave the horse-riding virgins who came to Deep Hollow all summer long in bikinis and cutoffs. It said: *You have no idea how hard it is to really ride a horse*. No translation needed. Under the Khejri, even as I felt myself fail the cavalry officer's regard, I cheered myself in the fact that I could read his eyes—that there existed an Esperanto of equestrians.

He walked to the pinto and cupped his hands beside his side to give me a leg up onto the stallion's back. The cavalry officer did not speak a word of English, his companion the rancher explained, but could he ever ride. "So can I," I said. In India, riding is a man's sport. There are no female cavalry officers, few female equestrians. Unlike in the United States, where the majority of amateur hunter/jumper riders are female, riding in India is an art born of war, and war is still a man's game. "Sure," the owner replied flatly.

As soon as my legs were around the horse, I felt a frisson, his pulse of readiness. Thousands of years of bred and boundless energy bounced from the horse's body to my own. I sat up straight and answered the horse's knotted excitement with my own stillness. He held. The cavalry officer looked up at me as he tightened the girth; it seemed to me that a whisper of new regard passed across his mustachioed face.

The officer mounted the black, and my horse fell in line as we stepped down the gravel drive. As we crunched out of sight, the rancher called out instructions to my guide in Hindi. My father spoke seven languages, but I speak only English and enough French to order a croissant. I had no idea what passed between them, except tacked at the end were two infuriating words, and in English: "No gallop!"

We rounded a bend and were soon out of sight. I trotted my horse up so I was knee to knee with the officer and spoke to him the few words we both shared. "Yes," I said. "Yes, gallop!"

The army officer flashed me a wicked grin, his beret flopping over his hairline as he swiveled in the saddle to turn toward me. Then he pivoted back to stare deep into the path ahead and pulled forward, so that all I saw were broad shoulders and a faint sweat stain seeping into his uniform shirt between his shoulder blades. I understood the word for what came next. "Hi-ya!" he screamed like a cowboy from a Western, at the top of his lungs. Underneath me the Marwari responded like he'd been loosed from a starting gate and simultaneously zapped with a lightning bolt. We were off. *Yes, gallop!*

How does a horse gallop? No one knew for certain until the 1870s, when a man named Eadweard Muybridge figured it out, and in doing so incidentally invented the motion picture. Oh, and he was also a murderer, though oddly, no one really seemed to mind that fact.

Muybridge, like my dad, had a thing for picking identities

that fit the moment. He was born Edward James Muggeridge, and when he was laying low in Panama after murdering his wife's lover in cold blood, he switched it to Eduardo Santiago.

In etchings and prints before the time of Muybridge's discovery, horses at a gallop appear to float over the ground with all four feet flung out fore and aft like a hobbyhorse—in fact, the name for the position. "Everyone knew horses didn't run that way, but it became the standard way of depicting a galloping horse in a painting," Phillip Prodger, a curator and one of the foremost Muybridge experts in the world, told me when I called him recently to understand the man. "In other words, a kind of galloping horse emoji that artists used again and again."

Although scientists had studied horses' gaits via their hoof-prints, no one before Muybridge's discovery knew for sure how a horse's footfalls were actually syncopated. And for some reason, the cadence of horse feet became a popular parlor debate of the day. The inquiry—if whether at any point in a horse's gallop all four hooves are off the ground—was picked up by the governor of California, a wealthy racehorse owner and industrialist named Leland Stanford, who went on to found Stanford University. He took the position that at some point in the horse's travels over the ground, the horse flies.

A self-taught photographer in the nascent days of the camera, Muybridge, a Briton, had traveled to San Francisco to chase California's bright light, ideal for his lens. There, the governor engaged Muybridge to answer his horse head-scratcher, equipping him with technicians to buy and fashion state-of-the-art lenses and shutters. He set Muybridge up with huge tracts of land on which to breeze racehorses before phalanxes of cameras. Crucially, he also footed his photographer's legal bill when Muybridge murdered his wife's lover, Major Harry Larkyns, in 1874. (Muybridge got off, even though the jury didn't dispute his guilt—they just felt he was justified. "The jury outraged the law and the facts and violated their oaths to set the assassin free because his wife was Larkyns' paramour," the *Russian River Flag,* a newspaper out of Healdsburg, California, reported with outrage after the trial verdict in 1875. "American juries everywhere are saying such murders are right.")

After laying low in Panama following the murder (and briefly becoming Eduardo), Muybridge, or whatever his name was, returned to California to perfect an intricate web of tripwires rigged to Stanford's cameras, a gallop-capturing contraption. In 1877, a Standardbred like Shader, named Occident, stampeded through the setup and proved Muybridge's benefactor right in sixteen photos: at the peak of the gallop, the animal is entirely aloft.

Those photos of Occident, projected rapidly in a series, became in essence the first moving picture. But the impact of capturing that image, that of a horse suspended in animation, had a profound cultural effect that went far beyond that. "Several of the positions which the horse assumes while in rapid motion were so comical as to excite the risibilities of the spectators," reads an 1878 article in the *Daily Alta California* newspaper about a show of the works. "By these lightning photographs, taken in less than the two-thousandth part of a second, Mr. Muybridge has completely exposed the fallacy of all previous ideas on the subject."

"It was a moment when human beings could finally do something beyond the limits of human vision—it was so profound," Prodger told me. "And once that had been broken, all kinds of things were possible."

The Marwari galloped at an impossible speed. Bunching and coiling underneath me, the brown and white stallion was a blur. I thought of the English crystal cups and saucers, swords and candelabras carried by animals like this over the breadth of India as I jangled in the tack, wondering how it—and how I—could possibly survive. Every few seconds the earth dropped beneath us, and the undulating trail through the pulverized rock of the quarry dipped and rose and rippled.

My horse's body billowed and constricted with every oscillation of the terrain. In front of us, the black horse's hooves flicked in and out of the dust flaring up behind the stallion and the officer before us. The air was loud with the rat-a-tat clatter of hooves on unmined marble, then a rumbling of distant thunder when they hit soft-packed dirt. The officer darted back a glance over his shoulder as we careened down a slope, and I realized he was checking if I was still on. By divine grace, somehow I was.

Then he went faster.

The breath knocked out of me as my stallion whipped through the hot air. He was going too fast for me to inhale, blasting into the air so that it compressed and felt almost solid, too thick to gasp. Tears from the wind of our flight streamed from my eyes. The flesh of my cheeks flattened against the pressure as the stallion tore into the atmosphere.

Then he went faster.

At the edge of a green-marble ravine, the officer pulled up, and my horse dropped from flight to an easy walk without the hint of a jolt. The man pulled a beat-up flask of water from a saddlebag and reached across the space between his perspiring black horse and my sweating painted one to pass it to me. I spilled some out on the fur to cool my horse's shoulders before taking a deep glug. The officer grinned at me then, and I saw in his face the democratizing effect of a shared passion. Worlds apart, in the saddles we were peers.

"'Gallop!'" he said through a robust guffaw, mimicking me, and the word echoed across the green marble mine as we laughed together. I peered through that heart-shaped gap in the world above my stallion's mane as we stood on the edge of the marble crater. We four were silent then, the officer and me full of the ecstasy of a horse's pure power, loaned to us mere mortals for moments we hope will never end. Holding his red beret to his head with one hand, the officer tipped back the canteen to his lips.

The water ran from both ends of his white mustache to rain on the exquisite horse beneath him. Neither minded at all.

NAZRANA

By the time the wheels of my 747 hit the tarmac back home, the obsession had begun. I wanted a Marwari. No, needed. No, pined for a Marwari, a horse whose ears looked like something Dr. Seuss dreamed up, wrapped up with the muscle and speed of a cheetah. That moment when the wind whipped up by that Indian horse pressed my cheeks to my bones stuck with me indelibly. Back in America, I had to find a Marwari of my own.

Turns out, I couldn't. Almost no one can. Marwaris are too special. The horses can't be exported from India, a de facto ban via a series of internecine bureaucratic rules designed to protect them, but which their champions say hurts the breed's vitality as a whole. The indigenous Marwaris are considered a precious natural resource in India, officially designated a threatened species by the Indian government, which has at different points impeded all export on grounds that their numbers are low and at risk of depletion. To export a Marwari now requires a written letter of consent from India's agricultural ministry, but for the past decade, it seems no one has been able to get such a letter. No one can get a Marwari outside of India.

And yet, under a slate sky threatening August rain, over a decade after that first sprint on the Marwari stallion, I stood at the edge of a pasture that sloped down to the saltwater glimmer of Cape Poge Bay. Here, 7,335 miles from Udaipur, on the island of Chappaquiddick off Martha's Vineyard, I found myself once again in the company of Marwaris. To be precise, one was softly slurping my hair.

Before me were six of the only Marwaris in America. How?

There are just over a dozen Marwaris in the United States, the lion's share owned by one woman. She is the savior of the breed, or its downfall, depending on whom you ask, but on her passion for the animals, everyone agrees. Francesca Kelly stood at the fence line beside me in jodhpurs loose at the hip and of white linen, her trousers literally from Jodhpur, India. Under the Englishwoman's broad straw hat, her salted hair escaped her braid, and on her wrists, silvery bangles jangled against one another every time she raised a hand to adjust the brim. She exuded the denuded elegance of a fur coat worn over denim, of a beach house shuttered for winter.

We had arrived at the farm on Chappaquiddick down the single paved road that bisects the tiny island plopped in Nantucket Sound. Chappy, as it is known, is a rural island with a single shop that proudly boasts its solitary status ("The Only Store On Chappy!" read a sign as we passed by the Chappy General Store and Service Station), but its humbleness is a bit of an act. Its shingle-style homes go for millions, and it is the refuge of osprey, cormorants, and the actor Meg Ryan. There's a bridge at the eastern end of the island, where on a July night in 1969, Massachusetts Senator Ted Kennedy careened a car over the side. Mary Jo Kopechne, a political aide who was with him, suffocated as the vehicle sank into Pocha Pond.

That's what most people know of the island anyway. Few who drive down the one road give any thought to the band of six horses who live on a rolling patch of sandy property, nipping grass spiced with salt blown in off the sea. But this tiny tribe is singularly exceptional: Kelly's horses are in essence the largest herd of Marwaris anywhere in the world outside the Indian subcontinent.

Kelly, the stepdaughter of a British diplomat, had lived exactly the kind of itinerantly glamorous childhood that her dowager-on-safari look that morning said she did. Her youth was spent under the shadows of the pyramids in Giza. Whereas I bundled into the family station wagon for weekends in the Hamptons, her family weekended in the desert. There, ex-pats pitched temporary compounds of bedouin tents, pleasure palaces of fabric. In place of my crawling Long Island Expressway, her childhood trek to a weekend away was on horseback, in the cool after sunset.

"There might be twenty to fifty people all riding at night in the dunes," she said later, sitting cross-legged on the floor of her home on Martha's Vineyard. The inside belied the conventional Vineyard exterior: a glittering, bull-sized statue of a dragon leered at us from the dining room. "It was magical to us because the sparks would fly off the flint or the quartz in the sand—you had the stars above and the sparks below from the horse's hooves." She unbraided her silvered hair and raked her fingers through it as she spoke. Before me, she seemed to morph from prim to siren with the memory of the hoof-falls, the tambour, the mirth. "And the wonderful sound of great happiness."

Grown, she found herself married and rich in London, a life as cushy as it was stilted. It was 1995, and, she said, pulling her knee to her chin, "the confines of my domestic landscape were chafing against my itinerant upbringing. I felt like I was drowning in convention."

Last-ditch efforts to save foundering marriages are not uncommon; some try for a child, others a blowout commitment ceremony. Her husband proposed a riding holiday to India. In a sense, it worked; she felt alive once more and fell deeply in love again—this time with a Marwari mare named Shanti.

She bought the little mare after the trip, unaware at the time of how jealously the Indian government guarded the animals. There was little precedent for exporting the horses, and every route seemed to be knotted in impenetrable bureaucracy; the mare was stuck in her native land. But visiting Shanti was a great excuse to stretch against the strictures of her London life, enabling Kelly to travel back to India regularly to check on her and attempt to secure the horse's passage to England. On trips back, she bought still more horses, with the help of Kanwar Raghuvendra Singh Dundlod, her initial safari guide, until she amassed a small India-based stable of the best-bred specimens she could find.

To convince the Indian authorities to open their ports and let her Marwaris go, Kelly became a one-woman advocate for the breed. Before she set foot in India, most breed registries for the Marwari had long been defunct, even though the breed has legendary status in the highest echelons of Indian culture. The heroism of horses in battle is immortalized in Indian ballads and literature. Chetak, a particularly heroic ancient steed who bravely attacked a war elephant in battle, still zips around India, in a sense: his name is emblazoned on a popular brand of motor scooter.

But British colonists viewed the creatures as inferior to their English thoroughbreds, and as their vise on the country tightened, the horses plummeted in popularity. The fortunes of Marwaris fell in concert with those of the upper castes privileged with their husbandry. These animals themselves are part of the caste system in a way. Breeding and rearing Marwaris before the British denigrated them was once considered elite. It was forbidden to people born through no fault of their own into the servitude and derision of the bottom tiers of Indian society. As the rajas were stripped of their sovereignty by the English invaders, their barns were emptied. Their regal stallions were castrated. The rajas' herds were sent to pull carts for the lowest castes, now cared for by the same people who once had been deemed too low to touch them. By the time Kelly found Shanti, the animal was seen by many as a native cur.

To elevate esteem for the animals in their homeland, Kelly and Dundlod, an Indian nobleman, set up Marwari horse shows and trials to display their usefulness. They helped create the Indigenous Horse Society of India in 1997 to establish a breed standard, studbook, and pedigree. It was painful for Kelly to see the often poor conditions of those animals bent to lives of labor and the corrupting effect of decades of indifferent breeding.

In 2000, she finally convinced the Indian government and the US Department of Agriculture to let her fly out six Marwaris. Their destination was the Vineyard; her husband had long planned to retire from London to the United States and she joined him once she found her horses their paddock by the sea.

Kelly's desire to resurrect the Marwaris, to send them galloping out of their homeland, bolstered the breed and reminded its countrymen of its value. But the goodwill soon began to fracture. A British woman inserting herself into the fate of this precious Indian natural resource, seeking to remove it from the country for her gain, insisting she knew what was best for them, smacked uncomfortably of something else: an equine-centric version of colonialism.

Soon after the horses arrived in America, she put a stallion, for which she had paid a few thousand dollars, up for sale, asking $50,000. It is a moderate sum for a quality stud, and perhaps even low for his rarity. But with horses in their native land going for a pittance, it was an astounding price tag to native breeders.

In response, the price of Marwaris in India spiked as local breeders clamored to follow suit. Fears that the limited number of these creatures would be ripped from the country for outside profit resulted in the threatened designation being slapped back on the breed, with exports grinding to a halt in 2004. Since Kelly's success, it has not been possible to export a single Marwari.

And yet in the years since, Kelly's herd has expanded. There on the island were not only some of the original imported horses, but new generations of Marwaris in America. Her foals, like the filly lipping my hair on Chappaquiddick? She was about four years old, from a new bloodline, and she has siblings. It was a summer day in 2018, and there had been no exports of Marwaris since 2004. When you do the math, it doesn't add up.

Marwaris may have been marooned in India, but vials of frozen Marwari semen have somehow winged their way through Air India flights to Chappy. They arrive in deliberately mislabeled climate-controlled packages at Kelly's island kingdom. She was oblique on the process, but the filly in the field was its illicit fruit—*and* the horse had a kid brother. For years Kelly has in essence been slipping rare stallion sperm past customs agents.

"So, um, how did you manage to smuggle, uh, horse semen?" I asked her.

"I'm not a smuggler!" Kelly said with a snort. "It really ought to be allowed."

In her living room in Martha's Vineyard, we poured through picture books of her exploits and videos of her dancing with the animals at exhibitions across India, her high cheekbones glittering in the glow of golden Indian jewelry. I sat sunken into her couch, which was festooned with pillows in rich Indian fabrics. On my lap was a bowl of sliced peaches doused in tart cherry juice that she had served to me. Kelly sat at my feet on the floor. She had spent the last hours spinning noble tales of why she has taken it upon herself to champion the breed—the pitfalls and the blind spots, the successes and the consternations.

There was one thing missing: the why. What had compelled Kelly, of all the people in this wide world, to adopt the Marwaris as her charge? *Why Francesca Kelly?*

She looked up at me plainly from her living room carpet. "You come up with tremendous adventures when you're engaged in tremendous duplicity," she said. My spoon with a sliver of peach came to a halt in midair.

I had taken the high-speed ferry from Manhattan to her seaside farm that weekend, whipping up the East River and blitzing along the Atlantic coastline to interview her. Kelly had readily agreed to meet when I called her, eager always to show off her equine treasures. I spent three days on the island with Kelly, her horses, and her granddaughter and daughter ("I'm a professional treasure hunter," her daughter, Amber, told me, handing me a business card that read, essentially, "professional treasure hunter").

Nights I holed up at an inn near Kelly's house owned by a crime novelist who gives writers and artists discounts on rooms and trades lodging for artwork. Each evening when Kelly dropped me off, I thumbed through the bookshelves full of past guests' books and pondered the bartered art on the walls, wondering when I'd get to the marrow of Kelly and her horses' story. To the truth.

The truth's name was Kanwar Raghuvendra Singh Dundlod. But everyone calls him Bonnie. It is a love affair that has spanned three continents, two decades, and her marriage. On that first safari in 1995, Kelly and the striking Indian rider to whose side Shanti, her alpha mare who couldn't stand being at the back of a hack line, had tugged her, had fallen madly in love.

And I get it. Kanwar Raghuvendra Singh Dundlod is *everything*.

Dundlod is of Indian nobility, or perhaps something closer to landed gentry; he's heir not only to a fort but an entire village that bears his family name. There in her Martha's Vineyard home, Kelly played videos of the pair on her laptop. There they were, dancing a lovers' tango on horseback as classical *Kathak* music thrilled around them, each horse step-stepping in time in its own stamping *nritta,* an Indian dance.

Dundlod looks the patrician part, with a white handlebar

mustache that curves down his cheeks. Amid his aristocratic circle—the privileged class entitled to the care of Marwaris—Kelly said she felt a homey familiarity. There was a grace in the formality in the Indian court. Over cups of chai, the horsemen she encountered had a studied decorousness of bearing, reminiscent of her blue-blooded upbringing. The strict social codes of the caste system were a nostalgic comfort, a subcontinental parallel to the upper-crust English sensibility of the world in which she was born. Plus, the foreign nobleman added a dash of hunky derring-do missing from Kelly's staid life. Dundlod is considered a pioneer of the horse safaris that now crisscross India. And he is credited with introducing India to elephant polo. Elephant. Polo.

When she found him, the connection was instant and intractable for Kelly, drawing her across the world to ride at his side for decades. Every trip to bolster the breed was a chance to be together. The horses were a side effect of their love. As they cooked up ways to be together, Dundlod and Kelly traveled the country to horse fairs, established breeder societies, and won and hosted races and demonstrations. Their passion for horses was tangled up in their passion for each other.

For Kelly, it was all bigger than both—bigger than the love for an animal or the love for a man. With Bonnie and the horses, she found a second chance to live a life that felt once again like that midnight march into the desert, full of sparks. "I would say it's a love story that encompasses horses," Kelly said. "What came first? The horses came first—but they are not the whole story."

*   *   *

The next day, at the edge of the Chappaquiddick paddock, I slipped off my riding boots and rested them next to a bale of hay, padding on bare feet to Kelly's black Marwari stallion tethered under a tree. He pricked his question mark ears at me. His name was Nazrana, and he was Kelly's foundation sire, the first stallion she managed to wrest from its country who arrived with her first imported herd in 2000.

At twenty-eight years old and a petite 15 hands high, Nazrana was still crested with testosterone-built brawn that thickened the typical Marwari "peacock neck," named for its arching, birdlike curve. "Like a raptor," Kelly said. Barefoot, I propped myself on the hay bale and climbed astride Nazrana bareback. I wiggled my toes into his warm fur, and we trotted off to the beach visible past the paddock's edge. Kelly trains her Marwaris in salt water. Week-old foals wade in at their dam's side. It is the first place they are mounted as young horses, she said, under her theory that the buoyancy of the bay lightens the load of the rider and reduces the strain of the experience on young equine minds and backs.

"Get him in the water!" Kelly screamed over her shoulder at me from astride her own mare. We had reached the sandy shore, and she had already plunged her horse into the dark blue, her young granddaughter in a frilly sundress riding double with her. The adorable child sat on the horse's shoulders, her little feet on either side of the withers, dabbling in the sunshiny sea. As Kelly exhorted me to lead her horse to the water and make it swim—with increasing volume and intensity that echoed down the shore—I suddenly understood how she bent the Indian government to her will to let her extract a precious natural resource.

Her formidable voice boomed across the water. Her Marwaris are practically seahorses, she said. Nazrana stepped a hoof off the shoreline gingerly into the wave, but he would go no farther. It was Samson from the Mounted Auxiliary all over again, but this time—the chilly bay lapping at his heels, hiding pinching crabs and stinging jellies—I understood the horse's refusal to budge.

"I can ride anything," I had glibly told Kelly when we first set off through the bramble to the beach, and while a lifetime of sitting on everything from city horses to the Montauk hack line bandits has made that largely true, that afternoon made a liar of me. Right in front of me, the threesome of sundressed child, behatted grandmother, and bay mare splashed along gamely. Nazrana refused to follow his mate into the surf. I begged him with my bare heels. He threw his four feet akimbo. Every ounce of elderly stallion might insisted on us staying landlubbers. I pressed my shoeless feet into his flanks and his peacock neck shot up. He tossed his curly ears. A horse could not have shook his head any more emphatically: "NO."

"I said I could *ride* anything," I called out across the water to Kelly, where she and the little girl were now hip-deep in waves, the mare swimming up to her neck. "I did not say I could *swim* anything!"

Sometimes horses are balky because they're ill-mannered grumps, but that's a far rarer thing in equines than it is in humans. Humans can be sulky, fickle, distemperate, and miserable, and so can horses—but in animals, it's rarely pure misanthropy. A sour disposition is often a window into the body. Peer through it, and more often than not, I have seen that resistance is not "no" but "ouch," not "I won't" but "I can't."

The pebbles on the beach squeaked underneath Nazrana's hooves as he sidled away from the water, and this time I let him. He was twenty-eight years old, healthy and fit but an elder statesman in horse years, and I heard his body. I walked up a hillock of dune and dismounted, ponying the stallion to a scrubby pine. But Kelly was still adamant I try out her seahorses and insisted that we switch mounts. As we traded the horses between us, Nazrana danced on the shells with a growing erection at the close proximity of one of his girlfriends. The old fella still had it.

I handed Kelly's granddaughter up to her on the stallion, then scrambled onto a driftwood log, swung up onto the mare, and headed into the sea. The mare was eager for the water the way a racehorse is for the track. As we went in deeper, I felt the uncanny swish of her legs losing purchase on the seafloor. I was riding a horse that was riding the sea. For the first time, I felt that maybe my mount understood what it was like for me, her rider, her body buoyed by the ocean just as her back carried me.

I glanced back at the shoreline. As I turned to look, Nazrana's hind end gave out from under him, taking Kelly and her granddaughter down to the sand as he fell. For a second, he floundered on the beach. Then with a heave, he was back up, the two riders hanging on as he scrambled upright from the seaweed and sand. He was sandy and startled but fine, and so were they. But Nazrana's swimming days seemed past him, and he had been trying to tell me so. It had not been "I won't." It was "I can't." Like all horses always, he knew better.

Back at the barn, I splashed the brine from the mare with fresh water from the hose, and we put Nazrana back in his paddock. His harem all around him, the sea far in the distance, the stallion was happy.

When I was tiny, I dreamed of marrying a horse. (Oh come on. Who hasn't?) In India, horses are part of many wedding ceremonies, but not in the way toddler me had hoped. *Baraat* is a wedding parade that puts New Orleans's second line marches to shame. A magnificently turbaned groom astride a pale mare, or sometimes in a carriage, leads a parade of family, friends, musicians, and revelers from his home to his bride's side.

The life rite was forbidden to the lowest castes in India, Dalits, or Untouchables, who make up about 20 percent of the country. And despite the fact that so-called untouchability was officially abolished in 1950, Dalits are still stoned or worse if they attempt Baraat. Nevertheless, increasing numbers do, part of a growing Dalit rights social movement, wearing motorcycle helmets instead of turbans as protection from the stones.

In 2018, a twenty-one-year-old farmer named Pradeep Rathod was murdered in Gujarat state in western India for daring to own a horse. Members of the upper warrior caste, Kshatriyas, warned him not to ride, and when he didn't listen, they rose up and slaughtered him. "My son's love for horses led to his murder," his father said in a police complaint, Agence France-Presse reported. They killed the horse too.

I think about those darkly internalized feelings with which I've always struggled—that sense that horses are for some other sort of human than me, that I'm an interloper in an equestrian life that was not made for my people. "The clientele is restricted to Gentiles," that 1930s flier from Deep Hollow Ranch read, and I sometimes still got that message nearly a century later in the exclusionary, upper-crust culture that imbues much of the sport I love.

But those are just the quiet, menacing whispers of culture and class that I hear so loudly because of my history; my family's story has left me with a seismographic sensitivity to such things. But what if the rules were hard and fast and cruelly enforced, and to love a horse was peril? What if I was told I was not good enough for a horse?

The same year Rathod was murdered, upper-caste villagers in the town of Nizampur offered a man named Sanjay Jatav a compromise: the Dalit could ride to his wedding, but only on the outskirts of town, not through its center, as their own sons had always done. He'd be safe from attackers that way, they said, according to reports, and he'd also bring them less shame. But that wasn't good enough to Jatav. He took them to court, petitioning the Allahabad High Court for his rights.

On his wedding day, 350 police officers marched through the middle of Nizampur. At their center, Jatav sat resplendent atop an ornate carriage drawn by a horse. The court had dismissed his case, but facing crushing public opinion, the town relented. "We fought against all odds to just earn respect. Dignity and respect for our community," he told a local television station.

Crowned in a red turban and swathed in marigolds, his smile on TV is irrepressible. He beamed at his bride between the extra-special ears of a Marwari horse.

Out at the edge of Cape Pogue Bay, the fruits of Kelly's transgressions graze on salty grass under those funny ears.

"Maybe I'm an outlaw by nature; I haven't really subscribed to any system," Kelly said that afternoon in her living room. It was the final minutes of the weekend I had spent with her, riding Nazrana, playing with her granddaughter, shopping for organic grapes together at a wildly overpriced farm stand. Our visit was ending. It was after she at last revealed her love affair with Dundlod, telling me of the hearts they both broke along the way. That admission had been followed in short order with her equally compunctionless explanation of how she smuggled Marwari semen to America, contravened international law, and created the illicit herd on the quiet island.

I sat with the bowl of peaches on my lap, untouched.

"My ex-husband used to say, 'You know, you cannot go through life thinking you're above the law.' And I said, 'Why not?' We should all be above the law, and above religion! The law was written for idiots and criminals, and religion was written for idiots and sheep. It's all a means of controlling people who don't have the ability to control themselves, who are in a sense lost."

I put down the spoon with a clink, and the slivers of peach floated in the tart cherry cordial, drowning in red—bitterness surrounding sweet that somehow improved it. She got divorced in 2009. Through Kelly's unhappiness, the Indian animals had been swirled into her world, and the Marwaris as a breed were enhanced. It was clear that her love for Marwaris had to do with something other than their proud necks, their noble history, or their fanciful ears. Something other than Bonnie. Or maybe even something other than horses at all.

"Sometimes it's, 'Why the hell am I doing this?'" she said, as if she'd heard me speak the question aloud. "And yet these horses have coalesced something in me, so many of my passions."

Folded around herself on the carpet at my feet, I sharply saw Kelly as that little girl galloping out from Cairo in the dark of the desert, where the only light was from the tiny flames each horse hoof struck in the sand. As we spoke, Dundlod was somewhere over the Atlantic Ocean, in the middle of another twenty-two-hour trip to be with their horses. He'd arrive on Martha's Vineyard the next morning, once more at her side.

In the velvet night of her childhood, there had been no paths to follow. Kelly had stopped at nothing to create a life that replicated that roadless, rule-free world, chasing those sparks.

"One thing leads to another in a passion, one story leads to another, one horse leads to another, one love leads to another, and so on and so forth," she said. Her eyes unlocked from mine to stare off somewhere far from that quiet living room.

סוס

[soōs] *Hebrew:* horse

The way my family approached my love affair with horses has underscored it as something that has never been of value. Yes, my father saw every rosette I collected as expectorating in the eye of the forces that tried to erase him. But that day at the Hampton Classic when he collected the prize for me was in fact the first and last horse show he ever attended. Riding wasn't opera, or a debate on the meaning of *Waiting for Godot*. Aside from the wry Freudian analysis he sometimes used to explain my fixation to his friends—"Young girls are obsessed with being in control of this huge thing between their legs," he'd scoff—it didn't interest him, so it wasn't interesting.

The truth is, my father and I became friends only once I was old enough to engage on subjects he found compelling; before that I have no memory of him. We never played, he never got down on a knee, and we never went to the park together. His version of a father was born in 1900s Poland and frozen in amber. There were no American games of catch, no daddy-daughter teatime.

My accomplishments with horses were not currency of value

to my high-pressure, high-power mother and father; horses weren't Harvard degrees or newspaper bylines. My accolades in the hunter ring didn't translate in their world of intellectuals and sophisticates and did little to augment their status in that stratum, and so they meant nothing at all.

I internalized my parents' perspective for years, leaving horse shows with a blue ribbon fluttering on my dashboard and the crushing sense, as the adrenaline seeped away, that my fixation on the sport was a dalliance, an embarrassment, something I should have gotten over long ago.

Then my father died. Dad fell at age eighty-four. He was attending a bris of an infant cousin at a synagogue on Long Island when he caught a toe on a thickly carpeted staircase. He fell headlong into the Jewish temple, shattering his right side as he did. He had been attending opera, solo as always, the week before, but immobilized after surgery, he swiftly spiraled into stagnation that drifted into his mind. Dementia ate away his memories, and while he could still count in all seven languages, he could no longer remember my name.

My fractured family cleaved further as Dad became babylike—small, frail, cared for by Beverly and her family. She had worked by then for my family for twenty-six years, and now my father was her charge. Bev and her sister Angie cycled through our house on Park Avenue, converting my childhood bedroom into his nursing home.

Each morning, they dressed him crisply as the psychiatrist he once was. He would slump in his wheelchair as they knotted silk neckties around his neck—the ones he once loved boasting about buying for three dollars on Eighty-Sixth Street from the guy selling them off the back of a truck. Beverly spoon-fed him cottage cheese that caught in his mustache and addressed him as Dr. Nir even as she changed him like a baby. Angie worked days, Beverly overnight, sleeping on a cot beside his bed. In the dark, she held my father's hand as he slept just as she once had mine.

As he diminished, I clung to him. I hired a musician who wheeled him over that Oriental carpet into our living room each week to serenade him. She played klezmer music on her accordion. My hope was he would hear the melodies of our people and find himself wherever it was he had gone. In those last moments, I tried to drag out every last thread of the complicated hero of my life. The only solace during that dark time as he dwindled was that Dad himself didn't know it. A crispy hot eggroll—cheap, like he always insisted we get—gave him as much joy as *Aida* by the end, and that, for me, was enough.

At the Jewish Museum in Berlin on a snow-shivery day not long after my father's fall, I stood in front of a small lithograph and felt myself rooted to the ground. I gazed at the fine lines of a stallion with a small white star on its forehead as it was paraded before top-hatted buyers. But it was the horse dealer in the 1869 picture by Friedrich Perlberg that ensnared me: the man wore equestrian boots, baggy breeches—and the ritual curled sidelocks, or *peyot*, of a religious Hasidic Jew. I stared at the lithograph for almost an hour.

In the sidelocked horse dealer, I saw the subversive cousin my father once told me about, the one who slipped pepper in horses and in his own way defeated his enemies. The eyes in the lithograph were impassive; to me, they hid a sly sparkle. But most of all, I stood there marveling that a Jewish horse dealer from the Old Country was real.

By then, Dad's thoughts erased by decline, it was too late to ask him the truth of the pepper tale. Time had stolen everyone from his generation as well, except one person: Aunt Buscia. She was 103 and came up to about my rib cage, and had once shocked me when she told me that the secret to living as long as she had was to "do nothing at all. Do not travel, do not eat new food, do not leave the house." I had petted the top of her head and reminded her she had been a nurse on several Israeli war fronts. "No! Is better to stay home," she said. Then she winked.

She lived in a bungalow in Tel Aviv with a lemon tree out back. When we visited every Passover from the time I was just a year old until I was fourteen, Buscia would let me twist off exactly one fruit. I would sit in a corner of her midcentury home and smell the heady scent of the sun-warmed flesh while the grown-ups chattered in Polish, Hebrew, and Yiddish.

I called Buscia there, and when she picked up, it took her a few moments on the phone to adjust to me speaking English. Even then, it took me a couple more minutes to convince her muddled mind that I was not a man with a woman's voice, while I stifled giggles. "Sir?" she kept saying. "But why do you sound like a woman?" Then it clicked for her who I was.

"Sarah? Who loves *sousim*?" she said, using the Hebrew word for horses. Over the years, my parents had sent her pictures of me jumping my horses. I remember that those photographs rested on her rosewood console table under a leaded glass lamp she had somehow spirited from the Old Country, which she luckily left before the war.

Excitedly, I told Buscia Dad's tale about the cousin, the pepper, and the rude location he stuck it. "Did we have horses in our family?" I asked at last.

There was a little bit of silence, as my words made their way across the Atlantic, the Mediterranean, and the Sea of Galilee, or maybe just through 103 years of thoughts.

"It is a nice fantasy," Buscia said. "But no horses."

After my father's funeral and the following seven days of shiva, my mom came with me to a horse show in the town of Saugerties in upstate New York. In search of normalcy, I had decided to go ahead with my plan to show Cover Story, a twenty-one-year-old thoroughbred mare I had been loaned for free on the condition I return her to the show horse she once had been.

When I was a child, my family had financed my obsession, augmented by the pittance I earned at the Red Horse and Deep Hollow. Unable to afford a horse on my own while I made my way in the world, I had cobbled together a riding life as an adult mostly by helping friends with their problem children animals or fostering freebie creatures. Story's body was sunken and her coat dull when she arrived at the barn I had begun riding at, about an hour's drive from the city in Tewksbury, New Jersey. By summer, she gleamed like polished marble, her thorough-bred inclination to fitness bursting her back into fighting form with just a little exercise and elbow grease. Story was so shiny and her neck so thick she looked ten years younger by August.

There were two shire horses in the yard of the bed and breakfast where my mother and I stayed in Saugerties, loaned to the innkeepers from a woman down the road to mow the grass. We sat on the flagstone patio and watched them in silence, their feathered heels hidden under the wildflowers, holding hands and sometimes sobbing for Dad.

The next morning at the horse show, I rode Story in low hunter classes, and while she may have looked like a teenager, her physicality betrayed her age. She struggled to cover the ground between the minuscule jumps, her stride compacted by creaky bones and the challenges of simply being old. I was reluctant to coax her, to push her harder, because underneath me was a good girl giving me all the effort she had. When we left the ring after winning an equitation class of tiny obstacles, a trainer at the ringside turned to me. He put his hand on Story's laboring side. "That's the best you'll ever do on the horse she is," he said. Any more felt unfair to Story; it was time for her to have an easier job. We headed back to the stables. We were done.

Lost in grief, I wandered between the blue-and-white-striped competition tents filled with world-class horses. Then I saw him. In a ring to my right was a jumper class, where a bright bay Dutch warmblood, as round as an apple and with the soft eye of a bunny, towed an inexperienced young girl around a course. Every movement of his soft and powerful body telegraphed, "I got this," to the kid wobbling around on his back. Those were the words I needed to fill the cavern gashed into my life by my father's death.

My father had been invincible, a warrior who, as he would say, defeated 80 million Germans as a child, part of the Polish resistance chucking Molotov cocktails up through the sewers of the ghetto. But death had defeated him. He was the fulcrum around which my existence spun, his trauma transmuting into my own, his goals and aspirations for his existence becoming mine. And he had stopped. The lilt of his seven-times-accented voice had been the soothing lullaby of my life, asssuring me that all odds could be defeated. His final golden curtain had come down, and the aria was over.

That horse was called Trendsetter. At that moment of grief and loss, my mother at last understood that for me, horses were not a hobby but a lodestar. It was one of the first moments I felt truly seen by her, and it represented a shift in our relationship that has held.

I was thirty years old, employed ever since I was thirteen, and long accustomed to making my own financial way. By that point, I was an investigative reporter for the *Times*, and later that year I would be named a finalist for the Pulitzer Prize. Yet a horse of my own was still far beyond what I could afford. But I was raw and weak, and grateful and ashamed, when my mother turned to me as Trendsetter galloped by, and said, "Dad would want you to have this." That day, she gave me some of her and my father's savings to purchase Trendy.

Since then, Mom has understood and supported my riding. Yes, she still mixes up a horse's halter with a dog's collar, but she knows what a lead change is now, even if she can't tell when I flub one. There with her between the rings of show jumpers bounding over poles in Saugerties, I felt as if I was once again that little me, wandering through the dark of our empty home. Only this time, I had at last found my mother, who wrapped me in her arms.

Trendsetter elevated a passion in me for the sport to an echelon I had only ever aspired to reach. With a horse as competent and capable as I was not, I could truly learn. With Trendy, I rode and won at a level I had only reached once before but never since—that single day at the Hampton Classic on Willow.

When Trendy aged and, like Story, needed an easier job, I leased him out, and the money he earned from the lease fee enabled me to purchase Bravo, a flashy green Hanoverian and thoroughbred crossbreed obsessed with grabbing hold of and zipping and unzipping any zipper he ever saw. Leasing Bravo to a sweet tween girl when he began to age helped me to afford Stellar, a powerful, dapple-gray Mecklenberg who makes even my biggest errors look glamorous. The money that Trendy and Bravo have brought in by my leasing them out will enable me to care for them in perpetuity when they retire. Trendy began an equine chain of biblical begats that has enabled my adult life with these creatures.

Sometimes I say my father sent Trendy to me from heaven.

TRENDSETTER

Exactly seventeen years to the month after Willow crashed to the ground, taking my vertebrae with her, so did Trendsetter. The setting was auspicious: it was 2016, at a competition held at the US Equestrian Team headquarters in Gladstone, New Jersey, on a sizzling August day. The circumstances weren't: trotting around, Trendy simply fell off his feet.

The country's Olympic horses are rarely at their official home in Gladstone, instead spending their careers jet-setting around the globe at qualifying competitions. In their absence, the 1911 stables hosts horse shows in a place that breathes with their legacy. Waiting for my heat to start at the competition that summer day, I slipped inside the headquarters. I stepped onto a cobbled aisle of red terrazzo tile, lined by stalls with oiled brass bars. Over the stall doors, I could almost see the specters of Olympic superstars, show jumping greats like ice-white Gem Twist and Stroller, the acrobatic pony who outjumped towering horses. I felt them there, swishing their ghost tails at ghost flies.

The USET headquarters gleams with the history of that horseflesh, glinting like the bite of a gold medal. In the empty

255

antique barn, I slipped up a creaking staircase at the heart of the building and found myself in a turreted clubhouse. The room was an octagon, a tower from which the New Jersey gentry who built Hamilton Farm, as it was originally called, could exercise dominion over its five thousand acres. The loudspeaker of the competition and the soft sounds of horses stalled below were muffled through a plush carpet. It was colored the brilliant green of a show jumping field. Dust skidded in the rays. They washed over moth-nipped antique Olympic uniforms and fraying rosettes of championships too numerous to count. That dignified setting would only serve to make my disaster all the more ungainly.

The loudspeaker crackled informing me that my division was about to start. I hustled back down the creaking stairs to the stadium outside, where the Olympic garret was backdrop to the competition. I mounted my corpulent Dutch warmblood, Trendy, and nudged him toward the ring. We were entering a flat class, a heat in which the horses are judged on the different gaits—walk, trot, and canter, no jumping.

While Trendy is a horse full of power, he is profoundly—pathologically, perhaps—disinclined to expend energy. At home at the barn I keep him at in Whitehouse, New Jersey, rather than spend hours strolling his paddock, something most horses enjoy, he regularly sidles up to the four-foot fence around his pasture and jumps it from a standstill. He then laboriously walks into the barn, straight to his stall, and attempts to put himself in for the night. He is an indoor cat of a horse. He is so keen to stay inside and potted-plant-still that I call him my ficus.

We had traveled to the show in a horse trailer that morning, Trendy cheerful, sipping from a bucket and pulling threads of hay from where it hung in a net before him. Pert as he was in the trailer, Trendy was adamantly sluggish in the August heat that afternoon. It was not the heat; it was the horse. Here in the shadow of an edifice to athletic glory, he was, as usual, bent on an unapologetic display of indolence.

At the judge's command, I somehow got him moving. The flat class was in a discipline called hunters, where horses are judged on their innate conformation to an ideal standard. There's skill to be sure; the horses must be piloted to perfection by their riders, but the horse either has that "it" factor or doesn't. Trendy has many darling qualities; he's safe the way a slow-moving city bus is safe, and handsome in a portly way like Marlon Brando in his later years. But he most certainly does not have that hunter "it." The horse show world has its own Mehitabel, and Trendy has as much of a shot at achieving her perfection as I did—which is to say we lose the hunter flat class. A lot.

Still, we try. I consider it sportswomanly to do so. I never compete to win; I compete to compete.

That loudspeaker chirped again as we chugged around at a canter, when the judge called for the twenty or so horses in the ring to decelerate to a walk. Without even a cue from me, clever Trendy braked with a relieved sigh. But something wasn't right.

Underneath me, I felt his right front hoof catch on the substrate of the riding ring. The arena dirt at Hamilton Farm is a costly, high-tech concoction of shreds of rubber and artificial sifted sand. It is designed to protect world-class sinews and tendons insured for millions. Putting a foot in that dirt for a horse feels something akin to strapping on a pair of Air Jordans. But packed with tacky shreds of bouncy elastic, it was suddenly quicksand for Trendy.

Trendy did not slow down so much as splat. He stumbled in the gummy dirt, pitched us forward, and then, midstagger, tripped on the sticky fibers again. All four legs crumpled, tangling limbs and hooves. And I was plummeting with him. Where there had been his big bulk, there was suddenly no horse beneath me as we fell together. As he pitched forward, the sudden inertia from canter to empty air catapulted me over his left shoulder. I hit the dirt and glimpsed his face just as it smashed into the ground, forehead first. We locked stares as we fell. His liquid eyes were wide open as his long muzzle dove into that expensive sand, all twelve hundred pounds of him cascading after.

With horror, my cheek in the dirt, body crumpled, I watched Trendy become a cartwheel of half a ton of horse, end-over-end, falling at speed. It was only seconds long, but I still see it frame by frame, like that morning with Guernsey when I was a toddler and he was a hero. On my side in the dust, I looked up from where I lay and saw Trendy's hind end falling toward me. As he flipped through the air, his haunches were hurtling down to where I was flattened on the ground. I covered my face and braced for the impact of his rear legs as they whipped down.

Then, they didn't. I opened my eyes and took in my breath. Beside me on the ground, Trendy lay still on his back, all four legs in the air.

When horses go down, they get up. They don't loll, they don't sprawl, and they certainly don't lay inverted and akimbo in the middle of an Olympians' arena—even the preternaturally lazy ones like Trendy. Within an hour of their birth, foals are on their feet, bracing under the sponge of a mother's tongue sopping up amnion, getting placenta and cord out of the way as womb-crumpled legs find their hooves. As prey animals, the instinct to be up and ready to run is ingrained in every fiber. Horses do not truly sleep standing up, as is a common myth, but most of their rest comes from dozing that way, dropping down for REM sleep only at brief intervals, when they feel secure. In a herd, only a few stretch out at a time, taking turns to stand sentry against the fanged night.

So when Trendy went down and stayed down, something shattered inside my soul. I leaped up, feeling things break and crackle inside my body but ignoring them, just as the same snaps in my spine didn't matter when all those years ago Willow's white face peered over me, dripping blood. Beside me lay Trendy, groaning softly, on his back.

I have seen one other horse not get up, fall and lie there, each hoof paddling desperately for safe ground. That horse was Hickstead, the gold medal Olympic mount of the Canadian rider Eric Lamaze. I had watched the pair power across thirteen jumps at an Italian competition on a November night in 2011, streamed from a Federation International Equest website onto my desk computer. It was a slow news night inside the *New York Times* building. Nocturnalist had wrapped, and I was working the thankless overnight shift given to cub reporters like me, too junior to have lives. The job was a catchall: monitor news wires for stories that broke after most (more senior) reporters went to bed. Mostly it meant checking in with the beleaguered meteorologist assigned to night phone duty at the National Weather Service for the latest in twisters and hurricanes. And if anyone important died, I'd get first crack at the obituary.

As he walked across the arena in Verona, Hickstead died. Still quavering from watching his death streamed online, I wrote his obituary.

"One of Canada's top Olympic athletes—a stallion named Hickstead—collapsed and died in front of hundreds of spectators while competing in Italy," I wrote in the *Times* article published the next day. "The horse had completed a nearly flawless round of show jumping at the sport's highest level, the grand prix, before he crashed to the ground under his rider and died in the ring, as fans gasped in horror in the grandstands." I continued, "Hickstead's dramatic death—the horse lay shaking on his back with hooves trembling in the air for several minutes— was not only witnessed by spectators at the stadium, but the event was being broadcast on TV and on the F.E.I.'s website. The unsettling video footage quickly boomeranged around the Internet." Right to my desk.

Now my own horse, Trendy, lay at my feet. I shook as I stood over him. It felt like I was seeing my own heart outside of my body, lying there in the dirt.

Trendsetter survived that tumble at the Equestrian Team headquarters. In fact, an emergency vet check that same afternoon he flipped over revealed he did not even get hurt. Later, I would discover I had crushed a vertebrae. Adrenaline kept me from feeling the pain. Accidents that have crackled my spine and shaken my core have at points given me pause about continuing this sport.

Very brief pause.

It's an extreme sport, often discounted as such because it is predominantly female. When I used to bartend nights after work at the ranch ended in the Hamptons, juicehead guys would sidle up to my bar. They'd order beers for themselves and "something girly and pink" for their girlfriends. I'd shake up cosmopolitans, a concoction of straight alcohol, just tinged pink with a drop of cranberry, and think about how the color and feminine connotation turned the drink—in fact, exponentially stiffer than a "manly" beer—into something "weak." Piloting huge animals at velocity is extreme, as wild as rock climbing, or motocross, and likely as treacherous. Sexism is the only reason I believe that riding isn't considered similarly badass.

It may be an extreme sport, but so is life, and as Trendy's tumble doing nothing more than slowly moving in a straight line showed me, there's no predicting whether you leave the arena in glory or on a stretcher. I'll never stop. I'm extreme too.

As I stood over Trendy, ring dust on my tongue and coating my helmet and showcoat, relief flooded through me when I realized he was breathing. So why was he not getting up?

"He's just stuck!" my trainer shouted. "On a flower pot!"

At the edge of the ring was a planter of wilting impatiens. It was solid, made of a wooden wine cask cut in half. Trendy, I realized, was wrapped entirely around it. The barrel was at his belly, and his fore and hind legs were on either side, almost like a person might hug a pillow to her stomach when sleeping. So enmeshed, he could not scramble his hooves and get them under him. When he tried to move a leg, his aluminum horseshoes kept glancing off the rounded planks of the planter. And my gelding couldn't flip on to his other flank either; his saddle had become a wedge between him and the ground, cantilevered so that he was unable to toss himself over it to his other side.

The term when prone horses get stuck against objects, or the wall of their stall, situations that those of us with opposable thumbs don't have to worry about, is *cast*. Horses injure themselves when they find themselves cast. Many flail, writhe, panic, stuck on the ground as their cells scream they should never be.

But true to ficus form, finding himself cast, Trendy instead gave up. More like a potted plant than he had ever been before, Trendy just lay there. He stayed totally still, legs akimbo to the sky, patiently waiting for humans to attend to his needs.

They came running from across the show grounds, that same fear in my heart quailing everyone around; the spectators were all equestrians who also knew horses never ever stay down. A mixture of rubberneckers and well-meaning riders swarmed Trendy, inadvertently upping the chances he would panic. Someone tried to grab his hind hoof and turn him over, but even the gentlest of creatures, upside down and thronged, could in their fear land a vicious blow. My trainer yelled for them to stop and then barked into the horde: "Get back. I've got this!" Out came a rope. We slung it around Trendy's hind fetlock, planted our feet and hauled. Over half a ton of horse followed the tug, flipping over the wedge of the saddle to his other side. Able to get a purchase on the ground at last, in a second he was standing. The crowd cheered.

Trendy tried to eat the impatiens.

My face burrowed into his neck, sobbing with relief, a strange thought followed as I tried to piece together the uncanny accident. I had fallen over his left shoulder as he thudded down, toward the center of the ring. Trendy had done a half-somersault, his head landing to point in the opposite direction we had been traveling. That made sense. But something else didn't: How had he gotten stuck on the wooden flowerpot to his right? How, when I was jettisoned with the trajectory of his falling body to the *left*, had he fallen to the exterior of the ring, to the *right*—the *opposite* direction? Why, when I braced for his huge hindquarters to come crashing down on me, had I been spared?

"I just watched your horse save your life," a woman in a showcoat and old-fashioned velvet helmet said to me as I sat on a bleacher. A paramedic was testing my reflexes after the crash, shining a light into my eyes as my pupils painfully contracted. "I was directly behind you in the flat class," she said as the EMT asked me to tell her what day of the week it was and who was the president of the United States.

"That horse was going to crush you; there was no way he was falling anywhere but on top of you," the stranger continued, as I shivered involuntarily. "And then, midair, he flipped himself. I've never seen anything like it. He torqued his whole body as he came crashing down and flung himself in the opposite direction—into that huge wooden barrel—to avoid you."

I slid my hand up Trendy's sweaty neck, where he hung his head over me, quietly observing my medical exam. There's a great risk of someone who spends so much of their lives with animals to imagine they are more than they are—that a horse performs for me because he loves me, that he speaks to me in a language others can't hear. Horse psychics prowl show grounds, capitalizing on our mystical, anthropomorphistic inclination. Once, unprompted, one such medium called to me as I walked by myself across a competition parking lot: "Those three horses thank you for protecting them; they need you and are grateful." I had both Trendy and Bravo by that point, and hours before, I had finalized the purchase of a sweet quarter horse–cross who I intended to lease out to help fund my riding. (It's almost like an equine Ponzi scheme. I call it my *pony-zi* scheme.) I was so startled that I took the woman's card. It was emblazoned with cats on a celestial background, and I *almost* forked over $115 for her to do a telepathy session with the quarter horse over the telephone.

Sometimes, however, in an effort to remain rational, I'm too staunch. I have to admit that there is something that transpires between two bodies bonded by physical contact that changes them both. It is an interspecies bridge that is perhaps parasitic, perhaps symbiotic, but it leaves the two halves greater than the whole. And as much as I try to avoid cat lady status when it comes to horses, I do deck Trendy's stall with a plastic menorah on Hanukkah because he's a Jew, just like his mom. But more than that, as that picture book, *The Girl Who Loved Wild Horses*, foretold, I've become a member of Trendy's herd, just as he is my family.

There on the bleachers, I realized I had known that Trendy had saved me, that he had hurled himself into a chain fence and a giant wooden drum, even though falling on me would have been softer (for one of us, that is). I had felt it there in the dust. I saw it as he lay there begging for help with stoic silence. I had glimpsed it as his liquid eyes locked with mine in the split seconds as he snowballed in the air, as his and my face smashed simultaneously into the ground. There was the knowledge in that time-gone-slow moment we shared that he wouldn't hurt me, or at least, he would try not to. I hadn't wanted to say it aloud, thinking myself made of more logical stuff.

"You'd better thank that horse," the woman said as she strode away.

I did. I do.

CADILLAC BOY

Trendy has a passport. It is red and stamped with "Konin-klijke Vereniging Warmbloed Paardenstamboek Nederland (KWPN)," the insignia of the Royal Warmblood Studbook of the Netherlands. It says he was born in 2000 on a farm in Holland, and in place of an unflattering passport mug shot like I have, there is a small diagram of a horse. His little star on his forehead is penciled in by hand. The onion-skin pages show he was shuttled around Europe as a young man, with stamps from veterinarians attesting to his vaccinations as he was raised and reared and trained across the Netherlands and Germany.

To be a Royal Dutch Warmblood is tough. Being born one is not enough to qualify for the studbook. They must earn the official designation, and the Netherlands coat of arms, a lion rampant. There are 450,000 horses in the Netherlands, according to the Dutch Equestrian Federation, and the KWPN studbook produces eleven thousand foals like Trendy a year. Babies must pass a series of tests after the age of three to be admitted to the KWPN studbook clique. There are thorough inspections, where horses are assessed based on their conformation, biomechanical movement, performance, and strength. The highest classification of a stud is "preferent." It is earned by few stallions, one of which is Nimmerdoor, one of the breed's most influential sires, whose name I found written in the faded lines of Trendy's passport: his grandpa. (Ambition, it seems, skips a few generations. Trendy is *preferent* only to me.)

On page 16 of Trendy's passport, is the final European stamp, October 4, 2009. That's the day he moved to America, traveling to New York the same way many Dutchmen do: in the cabin of a Boeing 747.

Just off the tarmac of Amsterdam's Schiphol Airport, outside a warehouse known as the Animal Hotel, I held a quivering gray gelding at the end of a rope. It was a warm spring day in 2018, and there was little to betray the warehouse's status as one of the largest animal transport hubs in Europe—except a trippy mural on one wall. The painting features an okapi queuing for airport security as a polar bear steers a forklift into a taxiing jet. I was there to understand the industry of horse importing, the same Rube Goldberg machine from a Dutch breeder to, say, a Manhattanite owner, that had years before sent Trendy across the Atlantic.

I had flown that same spring morning to Schiphol from John F. Kennedy International Airport in Queens, New York, with the expressed purpose of flying right back. With a horse. The return trip would be in the company of a phalanx of European horses being imported to America. Dutch warmbloods are a huge agricultural export of the Netherlands, after, of course, tulips and gouda, and the flying of horses from there, as well as across the rest of the world for international competition, is an equally big business. I was there to see horses fly.

First, in the waiting room at the Animal Hotel, I met a breeder named Jeroen Strik, killing time before the flight thumbing through a stack of magazines with names like *Mein Pferd*, or My Horse, in German. He had arrived early with his charge, Karieta Texel, a three-year-old Dutch warmblood mare, whom he had driven from a farm to the airport. When she descended from the horse trailer she was was slicked with sweat, foaming and dripping her anxiety all over the warehouse's concrete floor, the painted okapis looking on. Vibrating slightly, she stood with her front limbs spread, her back legs tight, like a tripod. Her hooves were braced against going anywhere—certainly not into the metal shipping container in front of her, into which a crew of handlers were about to box her up and ship her to her new home.

"Every horse is worth millions to the owner, from the children's ponies to the Olympic horses I dropped off last month," Strik said, as I took notes for an article on the whole affair that would later run in the *Times*. He folded up the magazine and headed out to load Karieta. Time to fly. "I travel with them each the same way."

Despite the misleading mural on the wall at the Animal Hotel, horses are not forklifted by polar bears into a cargo hold while penguins watch from the tarmac. Instead they are packed three abreast into shipping cubes, which are then hauled down the runway and hoisted up into the plane. The warehouse resounded with the metallic screech of these boxes of horses as they were towed through the building, rolling along on a raised platform like machine parts on an assembly line.

Just outside the hangar, a cherry red van splashed with pictures of horses vaulting the wings of airplanes pulled up. A man leaped out the cab and began pulling down ramps to reveal a quartet of horses huddled inside the box truck. He was as sweaty as Karieta. "We're late! These horses better not miss their jet," the man barked to no one in particular, and then to me: "You're doing nothing! Load him up!"

He stalked across the parking lot and shoved a lead into my hands. At the other end of it was a doe-eyed gray Hanoverian with a dog tag that said "Cadillac Boy" hanging from his halter. The six-year-old looked as bewildered as I felt. All around me, Strik and the handlers tugged other horses toward the back of the warehouse where silver shipping containers sat ready on a conveyor belt. Strik's jittery mare was not budging; it took a group of three men who surrounded Karieta's rear and gently shooed her up a ramp into the waiting box. Once locked inside, the machine boomed as it slid the massive metal box full of her and two other horses down its length, en route to the waiting 747. A fresh, empty container rolled down the assembly line toward me.

With my notebook and pen in hand, I realized the sweaty man had mistaken me for an airline employee inventorying the equine cargo, not a reporter desperate to remain an invisible observer. Most alarming, I realized as he stormed away, leaving me and Cadillac Boy in the rapidly emptying warehouse, the man thought I had the faintest idea of how to load a horse onto an airplane shipping container.

I was overcome with a horsewoman's pridefulness, the same

bravado that has gotten me bucked off and bolted away with each time a person has said, "You can't do that," and there was a horse involved. *Sure, I can load a horse onto a plane*, I thought, tucking my notebook into my waistband and my pen behind my ear for a better grip on the gray's tether. I don't know anything about airplanes, true, but *horses* I know. Slow moves, calm assertiveness, and confidence that soothe a horse's natural wariness should be the same whether you're putting a horse away in a stall or up in a jetliner.

I hoped.

In front of me was the conveyor belt. With a roar, the empty horsebox rolled the last inches onto the loading dock before me and stopped. A group of workers lowered a gangplank from it to the floor. In the semidark warehouse, the box looked utterly terrifying. Then I spotted a light at the end of the tunnel: at the front of the cube, like a beacon, hung a bag of bright hay.

Show horses learn how to travel; they zip along highways and byways from competition to barn in jouncing trailers, standing the entire time, no seat belts, bracing against the bends in the road, and some do the same in airplanes. There in the Animal Hotel, I learned what I believe is the secret to all that equine jet-setting: just as gummy worms help me survive a road trip, I believe horse travel is made possible by hay.

Timothy, alfalfa, clover, grass, hay: forage is a horse's everything, and they're always up for a snack. Scientists at Rutgers University Equine Science Center recently hung accelerometers and global positioning system trackers on horses' halters to study just how often they ate. The results were startling: pastured horses spend up to sixteen hours a day eating. "Are they always hungry, or is it just evolutionarily a 'see-food' diet?"

Dr. Karyn Malinowski, the founder and director of the center, said to me when I called her up later to understand the allure of hay. "Science just doesn't know yet."

I had met Dr. Malinowski back in 2010, far from her research farm in New Brunswick, New Jersey, where she sets horses to run on specially designed treadmills as scientists study their physiognomy and locomotion and where things like why some racehorses refuse to eat and become anorexic are puzzled over by researchers. I had attended one of her lectures during my society reporter days, held in an apartment overlooking Central Park. The apartment was owned by a hotel heiress, and everything was in haute rococo style. The fete was a high tea to learn about Dr. Malinowski's work, the apartment stuffed with equestrian-inclined women of a certain age, each throat strung with pearls. Someone had heard I liked horses and wrote for the *Times* and slipped me an invite.

We sat on settees that afternoon sipping Darjeeling as Dr. Malinowski held forth on a subject dear to that set: how not to age. She was bubbly and casual and prone to corny horse puns, something like, "Who needs a man when you have a horse? That's a more *stable* relationship," and I liked her immediately. Her research with equine growth hormone replacement therapy had resulted in an elderly mare who one morning in the lab did a full Dorian Gray, rippling with youth. "I thought I was going to win the Nobel Prize!" Dr. Malinowski said to the tittering crowd, who leaned forward on their settees to hear the secret. Unfortunately, the effects were temporary; it only worked as long as the horse received her daily dose. One month after the treatment ended the mare looked her age again. The crowd settled back. "Botox it is, then," I heard one matron whisper to another.

I had fished out Dr. Malinowski's business card after that flight from Amsterdam, almost a decade later, because I wondered if her studies had been able to answer a nagging question: Why do horses endlessly eat?

In part, it's protective, Dr. Malinowski told me when I rang her: constantly ingesting saliva keeps the horse's stomach lining from becoming so acidic it causes ulcers. Ulcers are the scourge of the stabled horse who lacks access to constant munchies. But what once may have been an evolutionary advantage—their lack of the ability to feel full that keeps them on the move—can be a risk for the stabled animal. "Every horse owner dreads that thought of a horse getting into the feed bin. Because that horse will stand there and clean those bins until they are empty—even if it's a year's supply," Dr. Malinowski said. "It will eat until it dies."

Compounding the danger is a quirk of the equine gut: horses can't throw up. "Humans and all other animals that have been studied, mammals, fish, they can regurgitate or—eww, vomit," she said. "So if there is a blockage, if there is a toxin, we have the ability to get rid of stuff very quickly." But horses have a strong muscle at the base of their esophagus; nothing can get past it. There are presumably many helpful evolutionary reasons for this, but the catch is that a tummy ache can turn deadly fast. Colic, which in a human most often causes nothing more than a grumpy baby, can be swiftly fatal for horses.

Yet why exactly horses are always hungry is something of a mystery. "Have they just through evolution learned to eat everything they can in preparation for a time when there might be nothing? Is it because it is a prey animal who did not have access to a full buffet of food 365 days a year?" Dr. Malinowski asked. "Science doesn't know." And that leaves another question still unanswered: gobbling constantly, a horse can go through twenty pounds of hay in a day. Do they ever become full?

Whatever the reasons, one useful side effect seems to be that as long as a horse has its nose in a flake of hay, even if the animal is going 80 mph down the highway or 500 knots through the sky, horses can be entirely at peace. On the tarmac with Cadillac Boy, I figured he was no exception.

I chirped for his attention and got his eye on the prize mounted at the far end of the narrow shipping container: the pile of alfalfa, ready for snacking. The bright clang of his hoof on steel didn't even register to him as we strode up the ramp and into the box. Cadillac Boy had hay-induced tunnel vision; he beelined right for the bale. Three echoing steps up the ramp, and he was all settled in the narrow standing stall, muzzle-deep in the hay. He was soon oblivious to all but masticating as I swiftly raised the bars at his chest and flank and jammed the bolts home, locking him in for the seven-hour journey.

"Why *de hel* are you up there in the cargo container?" the foreman of the warehouse barked up at me in his thick Dutch accent as I crouched to secure the last lock. I stood and stroked Cadillac Boy's forelock. "Oh nothing," I said, whipping out my notebook and waving it as an alibi. "I'm just here to observe!"

I boarded KLM's Boeing 747-400 COMBI the normal human way and headed to my seat in the very last row. Around me, passengers crammed into the aisles. They busied themselves with tucking away their carry-ons in overhead compartments or digging for their earphones, oblivious of the horses flying along with them. A vague *eau de stables* permeated the cabin, but if you hadn't known there were horses onboard, you wouldn't have noticed. Airplanes smell weird anyway. I peered out the window and watched as forklifts winched the last of the three steel boxes carrying Cadillac Boy, Karieta, and seven other horses between them up to the height of the plane. A hatch opened in the jetliner's side, and the operators slid the metal cubes laterally into the cabin—not into some below-decks cargo hold as I imagined. In the cabin, passengers fiddled with their screens. No one heard a lone, distant whinny.

The last row of the airplane were the seats reserved for grooms who travel with the horses. Visually, it appeared that I was sitting in the last seat of the plane. In fact, the cargo hold stretched from the wall behind the back of my seat to the 747's tail, accessed by a small portal in the partition. The door to get to the horses was almost camouflaged, and to the uninitiated, it appeared to be nothing more than a door to a lavatory.

It costs about $7,500 per horse, including quarantine stateside, to send them each from Schiphol to JFK. Next they are quarantined, where they are tested for pathogens like the tickborn *Equine piroplasmosis*, which abounds in other parts of the world and the United States is at pains to keep at bay. Piro was detected in one of Francesca Kelly's Marwari mares en route to America. Barred from entering the United States, the mare was exiled from every other country too. She lingered for months stateless in quarantine and was at risk of being put down until Kelly finagled asylum in Venezuela and a piro expert to cure her. Contagion free, Shymala was at last exported to Chappaquiddick. ("That year of quarantine did impact her equanimity somewhat, although maybe it's just her F-you personality," Kelly said. "I love her enormously.")

Quarantine stateside can mean different things for a horse

depending on its sex. For a castrated male, a gelding, it's a quick few days. For mares, who can carry sexually transmitted diseases, quarantine is a boring month in a US Department of Agriculture–approved holding facility.

For stallions, though, it's an orgy. They spend quarantine having sex with a harem of "test mares," animals kept by facilities for the purpose of culturing STDs, in particular contagious equine metritis, a venereal disease that causes bouts of infertility. It can be easily detected in female horses but not in male carriers. Test mares, which by law had to be tattooed with a "T" under their lip, are essentially labs for disease to be cultured. Some equestrians have criticized the method as antiquated when more modern methods for testing for the illness exist. Others worry what a month in an equine bordello will do for stallion morale.

"Not to be vulgar, but once they get a taste of the real thing, that is going to make the stallion way more aggressive," a trainer told me for a 2016 article I wrote on the subject in the *Times*. "It's harder for him to concentrate and keep his focus on the importance of the task at hand."

The test mares, who live on birth control, have a better attitude. "These girls do their job very well, and they love their job," a woman who runs the University of California's equine quarantine program told me. "They do like the boys."

The horses on my KLM flight that day were flown by a dedicated equine travel agency, the Dutta Corporation, which sends about six thousand equines a year all over the globe. Some are traveling to new homes; others are equine frequent fliers, international stars bouncing around competitions from Miami to Paris to Hong Kong, the animals sometimes spanning the entire globe within a week. Most important is to ensure that those horses never get jet lag, Tim Dutta, the founder and chief executive of the company, told me. Dutta worked with Olympic vets to calibrate the lighting and temperature of their onboard experience to mitigate the effects of jet lag and keep these athletes fresh.

Every horse person will remember the horrifying scene in the movie *International Velvet*, the sequel to *National Velvet* with Elizabeth Taylor. In it, a horse loses its marbles halfway over the Atlantic and must be euthanized—by bullet—lest he kick a hole in the plane. That doesn't happen in real life, Dutta assured me. Tranquilizers are at hand on every flight, and they work just fine.

In the final row of grooms' seats, I sat next to Cor Fafiani, a professional animal flight attendant for the past thirty-eight years. Cor whipped out his cell phone, flipping through pictures of animals he's flown. He flicked through snow leopards, horses, and a sea turtle sloshing around a giant tank as it crossed the Pacific by air. Tucked between our seats was a parcel containing the horses' passports. As we readied for takeoff, Cor ducked through the door in the wall to check each one like a TSA agent, verifying the documents matched the little tags hanging on their halters.

I was there, and thus might as well make myself useful, Cor figured, and so I was soon recruited to join him to check on the animals in the hold. In between shifts, Cor spun stories about all that had ever gone wrong in his decades of flying with animals—and how he saved the day. I didn't know whether to be unnerved or calmed. I settled on queasy. Once, ferrying an okapi from Jakarta, Indonesia, to Florida, the animal went ballistic, *International Velvet* style, in the hold, he told me. Its panic was matched only by that of its companion—a human zookeeper. Cor banished the keeper from the hold and turned out the lights. Okapi are jungle dwellers, he knew, and used to the dim of the forest canopy. In the homey dark, the okapi settled, Cor said, his chest puffing out somewhat. It went back to munching its hay. Hay. Of course.

As a newly minted groom for the Dutta Corporation that flight, my duties were to check on the horses, top up their water buckets from large drums every so often, and resupply them with all-important hay. I was soon to learn there was another duty: to soothe them.

I followed Cor through the little faux-lavatory door into a raw cave. Chunks of insulation padded the plane's ceiling. It was latticed by trusses, the exposed skeleton of the jetliner. The three silver boxes with the nine horses clustered in the center of the hold. The room was crisscrossed by tough nylon webbing that held each pallet of precious equines down. The tough nets turned the cargo hold into an obstacle course around the horse crates, forcing Cor and me to duck and weave over and under the ropes as we ferried the animals fresh hay every few hours.

As the air miles slid behind us, I became an equine version of the flight attendant with her cocktail trolley. Every twenty minutes, I headed into the hold. There I heaved up a large jug from the airplane floor and glugged water into their buckets, then hauled the buckets up to the height of their muzzles to wait for each to gently drink. Cor assigned the box containing Karieta and two others to me while he attended Cadillac Boy and the rest. By the time we crossed the English Channel, all in my box had settled and were munching; only one, a rough-hewn liver chestnut, even bothered to look up from his in-flight meal when we hit a jolt of turbulence.

Descent was another story. Communication with horses is beautiful because it's largely wordless and requires us intrinsically loquacious humans to learn to speak with something other than words. But when nearly four hundred tons of steel is dropping through the atmosphere, the inability to tell the horses inside it what the heck is happening to their equilibrium is really a problem.

I stood inside the horsebox with my three animals as the plane began its descent. They looked at me with faces that can only be described as full of horror. Not one could understand what was happening to their bodies as the plane plummeted, the reason for the butterflies in their stomachs as we changed altitude, or the popping in their ears. They didn't startle or jitter; instead, they froze, six eyes rolling white, six nostrils flaring and gulping in air.

"It's okay, it's okay. I won't let anything happen to you!" I said to them, embarrassed that within earshot of been-there-done-that Cor, I was chitchatting to animals that could not understand. Then I heard muffled sounds of Cor himself coming from the silver box next to me. He was speaking in Dutch, but we spoke a horseman's universal tongue: I knew he was saying the same soothing things to his brood.

As the jetliner tipped earthward, the three horses squared their feet and shot their heads up for balance. They were a bay filly with a blaze, the liver chestnut, and Karieta, chocolate brown. Grooms do not have to be strapped into their seats for landing. Our job is to make sure the horses stay secured in theirs, as it were. I stood in the horsebox, braced against its walls as the plane tipped down, steadying myself with my grip on the horses' halters. As we descended, I reached from nose to nose, scratching here and there, palming handfuls of hay to their lips, and wishing I had three arms to soothe every muzzle simultaneously. Then, with a grinding wheeze, the landing gear deployed.

As one, the three horses bowed their heads over their stalls and shoved them into my chest. Each leaned on me with the full weight of its body, all pressing their foreheads into my torso like they were little children hiding in a mother's skirts. More rightly, it was the way a foal will huddle up under her dam in a moment of uncertainty, press into the warm fold between the inner thigh and the udder, and wish for the big scary world to disappear. And it does. There in the air, with a trio of animals pressed into me for solace, I held their heads and leaned back into the horses as we fell through the sky, assuring them that in my arms they were safe.

It was a moment I will never forget; when three frightened horses looked at me and called me tribe.

When the wheels hit the runway, the trio bunched, pushing into me more deeply. The plane slowed to a gentle taxi, and the four of us separated, hay on my sweater the only reminder of our clinch. In a few minutes, the horses went back to peering around with less terror and something closer to wonderment. Karieta reached out to sniff me, nibbled a strand of alfalfa off my top, and buried her nose back in her hay.

"Welcome," I said to her, "to New York."

TANGO

Baying hounds answered the clarion call of the huntsman's brass horn. It sounded through the trees of North Salem, New York, on a November morning in 2018; my horse's body grew electric underneath me with anticipation of the hunt about to thunder off. I shoved down my guilt about the little tuft of red fox cowering somewhere in the woods, lamely justifying my presence among the Golden's Bridge Hounds hunt club that dawn with the fact that I was there in the name of journalistic inquiry.

I was not here to hunt foxes, I told myself, though I took pains to look the part: I had knotted a stock tie around my throat, affixed with a millimeter-wide gold pin I had borrowed, after my host admonished me that the centimeter-wide pin I had brought was far too flashy. I had bought a wool jacket, a brand favored by hunters called a Melton, for the occasion at Manhattan Saddlery, the last remaining saddler in Manhattan, tucked since 1912 on East Twenty-Fourth Street when it began as Miller's Harness Company. Underneath was a tweed vest my host had loaned to me; it had been her husband's. On each button was a tiny scampering fox. I held the reins of my mount, who was pawing in anticipation, in fawn gloves.

My quarry was the real Juliet Faust.

Before I recounted the story of Juliet in this book, the stranger who had given me Adonis, I wanted to verify that I had recalled it correctly, so I called her nearly fifteen years after that first conversation. She picked up on the first ring with a smoke-braised, "Hello."

Now seventy, Juliet and I chatted as she sat in her new home far from Tennessee, in the heart of horse country in North Salem, New York, where she had been master of the Woodbury Litchfield Hills Foxhounds. She had just returned from a ball at the Sherry-Netherland hotel in Manhattan, a gala for the region's huntsmen and huntswomen that was an echelon above black tie: "Men will be in scarlet," the invitation read.

Despite the fact that when I was in my twenties she gave me, a stranger, her horse, I had still never met her in person. I wanted to know more about her. I wanted to talk.

Juliet just wanted to ride.

"You'll come up and we will go fox hunting, this weekend, yes?" she said after about four minutes of conversation, in that tone of a person for whom that "yes?" is more a conversational convention thrown in at the end of a sentence than it is a question. I stammered out the semblance of a no: I didn't have a hunt horse, I didn't know how to fox hunt, and, crucially, I didn't know her. I was a stranger who had called her after a long absence, no more known to her than I was that day I called her out of the blue from the parking lot at the stables in Amagansett.

"Then that settles it," she said as I struggled to figure out what part of my demurral had settled anything. "I'll get you a horse, and I'll see you on Saturday." The line went dead.

Tally-ho!

When the entire premise of a sporting activity is based around haranguing a russet ball of fluff; running it ragged before the seeking wet noses of a horde of hound dogs; plugging its croft so it can't burrow an escape; pounding on its heels across hill and dale and prairie and slope and marsh and meadow and orchard and briar, until you finally catch it and let the dogs vivisect it with their fangs—you could forgive me for hazarding that the sport is not the ideal hobby for me, a life-long vegetarian.

That's fox hunting, the bloody nut around which a genteel sport has been long affected, prettied up in scarlet coats and canary-yellow vests and crisp white cravats stuck through with golden brooches. In 2004, hunting foxes with dogs was banned in its homeland, the United Kingdom. Only a simulacrum called "drag hunting," where foxy scent is smeared across a landscape for dogs to track, remains legal. In North America, where 147 hunts are registered with the Masters of Foxhounds Association of America, it seems to me you can do what you like to the creatures.

The bugle burst through the chill air, all around me were the glitterati of North Salem, a town where the average income of a family is four times that of the country's average, according to the most recent US Census. The roster of the huntsmen and huntswomen around me, astride horses downy with fall fur, could have been torn from the pages of the Fortune 500. In

England (the crude joke goes), reckless, wild, and invariably drunk fox hunting is how the aristocracy are culled. The United States lacks the peerage, of course, but nevertheless, the fox hunting demographic here across the pond is so upper-crusty you could butter it.

That feeling I've battled for so long, the interloper astride, pierced through me. My horse, a black-and-white tobiano named Tango I had borrowed for the occasion, shifted sideways, eager to be off, and my field boot smacked against the polished leather of one of the rider's boots beside me. I shifted Tango out of range, aghast. It was the master of the hunt, and I felt like I had broadsided him.

Hierarchies are a favorite among haves, whether it's a mean girl in middle school ranking her peons, or a bunch of people from Westchester, New York, aping English aristocrats. Leaders of the horsey hunt pack are dubbed *masters*; they choose the day's terrain and keep the herd on the heels of the *whip*. That's the elite member of the hunt whose job it is to ride out with the dogs (gasp! I mean *hounds*) and keep the canines in line and tracking their prey. Masters also dole out favors, honoring long-standing members with "colors," that is, the right to swap out the simple black woolen hunt coat for a scarlet one, a sign you've achieved stature in the hunt. Think of it like being knighted for chasing a lot of innocent foxes.

Jostling one another is not done. Nor is cutting someone off as you careen through the woods; it's as rude as snaking a wave from another surfer. Maintaining your rank out in the chase is easier said than done, because horses traveling in a herd, as hunting parties do, get wild. As drunk on stirrup cup as the hunters may be, the animals beneath them get heady on a cocktail of instinct and joy, panic and pleasure. Breakneck through the woods, horses' hearts grow bolder. Powered by the herd animal's desperation to not be left behind, they leap tree trunks and stone walls that, alone in the forest, they might otherwise quail at, just because the horse in front of them did—and that one only because he was following another.

The only horse who must truly be bold is the master's out front, playing pied piper to his herd behind. For the rest of the riders, on horses juiced by adrenaline and endorphins, it's easy to get run away with when the horse touches the top of his power and, thrilled at the sensation, goes even further. My horse was raring to go, but we were bound by Juliet's command before we left: no matter how much of a freight train your animal becomes, do not dare to pass the master.

The hounds blasted off, and the hunting party trotted after them, threading through wooded trails that poured down a hill into farmyard. They had not yet picked up a scent, so the hunt was leisurely at first. The horses bunched together, and we stepped between paddocks when on either side, a herd of llamas popped their heads from grazing. Erect, they stood like puffy question marks in the field.

Then they advanced.

Tango hadn't blinked when a lost bitch threaded between his legs and had only glanced at a deer and her fawn on the side of the trail earlier, but these animals were too much for him. From that point on, Tango was committed to overtaking everything in his path, just to make sure the freaky llamas were as far behind him as possible.

The scarlet of the masters was stop sign red to me—do not pass—but all horses have dichromatic vision: green and blue are differentiable, but they can't see red. (More to the point: what does a horse know or care of absurd human conventions ludicrously imposed on a band of animals galloping through the woods?) My stomach knotted with the fear I would breach decorum, hauled by Tango into the front where everyone in the hunt would see I truly did not belong. My biceps ached at my failed attempts to tug him back into steerage class.

About a half hour of trotting aimlessly with no vixen on the wind (as I hauled on my horse to keep him in line), the band of about twenty paused on a hillock to let the hounds sniff and yip in the undergrowth. Juliet, made less bold a hunter by her age, was soon out of sight behind me, preferring to plod along in what is called Second Flight—riders who follow the rollicking First Flight of the hunt at a slower pace from afar.

As we rested, a man steered his horse toward me and handed over a crystal flagon in the shape of a crescent horn. He was a professional horse trainer, who was there babysitting a hedge fund billionaire who was festooned in colors but otherwise dubious as a rider. He plopped along with us, hunched in the saddle and hanging on for dear life. At least he had the good sense to bring a minder.

The trainer held the sterling stopper as I tipped the brandy back down my throat and passed it back between our horses. He proceeded down the line. At the next stop, it was a silver flask pulled out of a scarlet breast pocket and schnapps. At the next, it was bourbon, I think. We had been out for hours, and by that point, I couldn't recall what flask it was in or who handed it to me to guzzle.

Then the hounds bayed, the bugle called, and we were off

after some woebegone coyote (they'll do when foxes can't be found). We crashed over rocks and down hillsides and slalomed between conifers. Tango bounded over split rail fences on the heels of his compatriots, sluiced through mud, and whipped it back on my face. We did not slow for sheer drops, just let the horses find their footing and scrabble down. At the base of a hill, one mare pulled up bloodied at the pastern, and I called out to her rider that she was hurt. He dismounted, blotted her heel, and they walked off, the man leading her home through the trees on foot. We galloped on.

Before me was a mammoth stone wall. A few others were already waiting on the far side with the master, who had spurred his horse across and inspired the rest. I readied Tango for our turn, and we galloped toward it when the ungainly Forbes titan in red whipped in front of us, the trainer unable to control his ward. When the horse got to the stone wall, he did not jump it (for the man's sake, thank God) but veered sideways and ducked through some brush and around it. Tango was flabbergasted: Was he to follow that horse's lead and duck around? Or meet the horses standing on the other side of the fence, waiting for him to jump? I didn't know either. Befuddled, as we cantered to the base of the wall, we did neither. Just shy of the mossy rocks, we crashed to a halt. The hunting party gasped. And looked away.

For two years, I trained Adonis, the horse that Juliet, a stranger, sent to me. He was a hunt horse who knew what was needed for the field: how to stop and how to go, and not much else. By the end, he was a show horse, rippling with muscle and giving command performances at all gaits and paces. Then Juliet called. She wanted him back, a furred bit of the comfort of Tennessee as she set up a brand-new single life in the town she had moved to, North Salem.

She had joined a hunt, she explained, and founded another and was out riding seven days a week, rain, sleet, snow, or shine. I returned him gratefully, and she sent me *another* horse, a younger red-headed full brother of Adonis whom I named Aslan, who was too much of a toddler for her. Aslan turned out to have a neurological issue and became the beloved backyard pet of a barn-mate of mine.

But Juliet's ex-husband never had horses again, she told me the day we at last met at the hunt. After she left him for North Salem, he sold off thirty years of gear and any animals she didn't pawn off in a fire sale as he recovered.

"It was a beautiful life having my own hundred-acre farm and raising these babies with my husband, but life changes; you go along with it," Juliet said as we caught up after all those years over champagne at the breakfast after the hunt. More than a decade had passed since her heartbreak; it was no longer raw like that day in Amagansett.

Beside her, I felt excruciatingly chic. We had switched from our formal black wool hunt coats into tweed breakfast coats, as decorum required. My cravat popped over the lapel of the jacket I had borrowed from her, herringbone in seafoam green, atop the loaner copper-colored Tattersall vest that had been her husband's.

"That's the other thing I couldn't understand about my husband," she said, taking a sip of her champagne beside me.

"Even if I couldn't ride, I would want to have a horse in my front yard, someone to brush and take care of and to just see. To watch them run."

I didn't tell her what I thought then: he wasn't a horse person. And he wasn't your person.

I didn't need to. I was interviewing her in a mansion owned by one of the Golden's Bridge huntsmen, and her cheeks glowed with the just-finished ride. The carpet was powder blue, and it offset the canary yellow walls splashed with pink flowers, studded with woodcuts and oil paintings of hunters in pursuit. Around her were huntsmen and huntswomen with whom she'd forged a new life, at a gallop. She smoked cigarettes, setting each in a little holder before she put it to her lips, like a forties femme fatale. Seventy years old now, she still rode every day, sometimes twice. She chatted on about her new young horse, Gumdrop, and her plans to train him to stomach the wild abandon of the hunt. She had sold Adonis to a North Salem neighbor, who still keeps him in his dotage in her backyard. I asked her if she has remarried or was dating.

Juliet laughed.

"I have horses," she said. "How do you compete with horses?"

The mossy wall still in front of me out there on the hunt, Tango and I gathered our wits after we slammed to a stop. Somehow I had stayed on. Tango was confused but unscathed. But internally, I was in pain: I seethed with shame at being unable to execute so publicly in front of snooty strangers.

I gathered Tango's reins and trotted him around that same

wall I could not surmount, to join with the riders waiting for me, massed on the hill. As they headed off again, I turned Tango toward the tail of the party, to trot at the back with Juliet, my hostess in Second Flight. Something about the pomp, the circumstance, the wealth, and all the unwritten rules I could feel braided invisibly all through the experience had reverted me to the girl I was in the pony ring. Like the girl I was, I was waiting to be found out that I wasn't supposed to be there. At the base of that unjumped wall, I felt they had.

Older now, I knew that what I felt was a type of generational trauma, *belonging and not belonging* osmoted from my dad and my mom. For years now, I had kept that feeling stifled. And yet, like the fox, it may have been out of sight, but the scent of it was always in the air, somewhere in the distance, ready to close in. The sensation was overwhelming. Waves of anxiety told me to peel off into the forest and gallop to the safety of home.

Then another red coat approached me as we stopped in a gully. He did not deliver booze but something more intoxicating: "The master saw what happened and would like to invite you to ride beside him at the head of the hunt."

Wordlessly, I followed him past the other horses and riders to the head of the herd. The master had witnessed my minidisaster and seen, as I could not, that it was not my fault. *Mastery* was my father's catchphrase. It was his demand that I never hide as he did. He believed that what all of us are searching for is a sense of mastery, to own our place in the world.

I was startled as the hunt halted to let Tango and me step to the front, to realize they had judged my riding and determined I belonged—not just in the hunt but at its apex. I shook hands with the master, and we were off, side by side through the close-cropped hay fields, down shale hills, and over huge rock piles. Some appeared four feet tall on takeoff, but on the landing side, the earth dropped off far lower, so that your stomach fluttered as you descended through thin air. Ears pricked and riotous, Tango was in charge. Vaulting over the felled trees and fence lines, I felt I had vaulted through the stiff hierarchy; this pageant of aristocracy had coronated me, placed a sash across my shoulders that seemed to flutter as we galloped: *you belong!*

On a ridge, I halted beside the master as he pulled us up to listen for the hounds. They'd lost the scent and were ambling aimlessly through the woods, getting distracted by squirrels. I had agreed to go on the hunt mostly because Juliet swore they catch a fox only about one time a year, and the dazed hounds made me hopeful that today would not be their lucky strike.

A fine snow started to fall. The horses steamed and breathed in a stand of trees, close to one another for comfort, as the call and growl of the hounds just out of sight played through the forest. My misgivings abated, or perhaps it was all the glugs of booze, and I felt found in the wilderness, as horses so often let me feel.

"Why does your saddle say 'Nir' on it?" It was the bumbling hedge funder. He was pointing to my name engraved on a brass plaque on the back, or cantle, of my saddle. "Are you related to Yehuda Nir?"

I swiveled on the pinto's back. "He's my father," I said. The man replied, "I read his book in the nineties." There, in the sprinkling snow, he recounted a scene from my father's memoir. As a teenager in hiding, with the false identity of a Polish Catholic, my father had been interred in a German labor camp. When it was time for a mass shower, presided over by Nazi guards, he feared that naked, his Jewish identity would be exposed—he was circumcised. But the Nazis had made the mistake of positioning female guards for the showers. The undersexed prisoners went wild, whipping towels at the women and hollering; the mayhem provided a distraction for my father to duck in and out unnoticed, his Jewish identity still hidden, his life spared.

"At least every few months, I still think of that scene," the man said, tugging the reins of his horse, who had begun to fidget with the stillness. "I think of your father. The survivor."

It was silent, but I heard the roar of history and time, of belonging and not belonging, vibrating the air. Here in this closed-off world, an outsider had led the hunt and a Holocaust survivor had ridden alongside her, perched on the back of my saddle. I thought about the kick my father would have gotten, living in the minds of world-beaters, a *vincitor* galloping along with them on their tony pursuits.

Dad felt resurrected then, as if I would see him through the trees in scarlet at any moment. I could hear him among the whispering birch: "Mastery." I understood what he meant then: that happiness is not stuff or accolades but mastering a craft, a skill, a world, the self. The feeling of understanding life and bringing it, like a hound, to heel.

In the silent, snow-padded woods, I felt like my father was there with me. Masters riding side by side.

ACKNOWLEDGMENTS

I consider *Horse Crazy* itself an acknowledgment, two hundred and ninety pages long.

What you've just read is my tribute to the horses and humans who have let me into their barns and hearts, fired my passion and stoked my intellect, or just let me bury in their arms or fur. Each person and animal named here has shaped me. And they have graced me beyond measure by sharing their expertise or telling their tale.

I often say that the act of interviewing someone should not be confused with a conversation. It is instead an act of giving on the part of the source, and as a journalist I never let myself forget that each story is gift. This book is my thank-you for that extraordinary gift.

There are people not in these pages who have left indelible marks on my life that can't be captured in print (or they just aren't particularly horsey; I do spend time without equines— sometimes). They are my herd, my beloveds, who know they're in fact inscribed here in every line.

To Philippa Brophy, my agent, who told me my story would

find me, and Sean Manning, my editor, who sought me out because of my investigative journalism and didn't blink when I said I wanted to write about ponies.

And Paula Trachtman, whose daughter Amy lives on in all beautiful words.

ABOUT THE AUTHOR

Sarah Maslin Nir is a staff reporter for the *New York Times*. She was a finalist for the 2016 Pulitzer Prize for "Unvarnished," her investigation into New York City's nail salon industry that uncovered the exploitative labor practices and health issues manicurists face. Before becoming a staff reporter, Nir was the *Times*' nightlife columnist, covering two hundred and fifty-two parties in eighteen months. She has reported for the *Times* from around the world, from West Africa and the Alaskan wilderness, to post-earthquake Haiti and wildfire-ravaged California. Nir earned a master's at the Columbia University Graduate School of Journalism and graduated from Columbia University. She loves horses.